*New Perspectives on
Economic Growth
and Technological
Innovation*

New Perspectives on Economic Growth and Technological Innovation

F. M. Scherer

BRITISH–NORTH AMERICAN COMMITTEE
BROOKINGS INSTITUTION PRESS
Washington, D.C.

ABOUT BROOKINGS

The Brookings Institution is a private nonprofit organization devoted to research, education, and publication on important issues of domestic and foreign policy. Its principal purpose is to bring knowledge to bear on current and emerging policy problems. The Institution maintains a position of neutrality on issues of public policy. Interpretations or conclusions in Brookings publications should be understood to be solely those of the authors.

THE BROOKINGS INSTITUTION

1775 Massachusetts Avenue, N.W.
Washington, D.C. 20036
www.brookings.edu

Library of Congress Cataloging-in-Publication data
Scherer, F. M. (Frederic M.)
    New perspectives on economic growth and technological
innovation / F. M. Scherer.
      p.   cm.
    Includes bibliographical references and index.
    ISBN 0-8157-7794-9 (cloth : alk. paper)
    ISBN 0-8157-7795-7 (pbk. : alk. paper)
    1. Economic development—History. 2. Technological innovations—
Economic aspects—History. I. Title.
    HD78 .S34 1999   99-6202
    338.9dc21   CIP

9 8 7 6 5 4 3 2 1

The paper used in this publication meets the minimum requirements of the American National Standard for Information Sciences—Permanence of Paper for Printed Library Materials, ANSI Z39.48-1984.

Typeset in Palatino

Composition by Harlowe Typography
Cottage City, Maryland

Printed by R. R. Donnelley and Sons
Harrisonburg, Virginia

Preface

Following three turbulent decades, there are signs that vigorous economic growth is attaining renewed momentum in North America and Europe. Many of the Asian "tiger" nations, on the other hand, have suffered catastrophic reverses in what had previously been extraordinarily successful growth experiences. Because growth in the output of goods and services is almost synonymous with increases in the material well-being of nations' citizens, the sources of economic growth and the prospects for sustaining it through the twenty-first century are matters of extraordinary interest to business, trade union, and government leaders alike.

After struggling with the question for more than two centuries, economists have begun to advance new theories of how economic growth occurs and what conditions its sustenance demands. To explore these new perspectives, the British–North American Committee (BNAC) asked one of its members, Professor F. M. Scherer of the John F. Kennedy School of Government, Harvard University, to prepare an analysis of the growth phenomenon and economists' efforts to elucidate it. This book, whose core chapters were discussed by BNAC members at three meetings in 1997 and 1998, is the result.

Throughout much of the nineteenth century and less consistently in the twentieth, economists feared that economic growth would be choked off by diminishing returns—at first, on the finite endowments of land suitable for feeding growing populations and later on scarce energy and metallic mineral resources. These pessimistic prognoses have been upset repeatedly by the implementation of innumerable technological innovations. Technological advances have also made it possible to offer the world's consumers an array of new goods and services far beyond anything that could have been imagined two centuries earlier. Scherer traces how economists came to recognize the key role that technological innovation plays in economic growth and the ways through which advances in knowledge and technology feed upon themselves in a kind of virtuous spiral.

As we enter the twenty-first century, questions arise whether the technology-led economic growth of the past two centuries can be sustained. Addressing those questions, this book focuses on two crucial inputs into the processes of technological advance— capital investment and highly qualified scientists and engineers (embodying what economists call "human capital"). It characterizes the risks that investors in technological innovations must surmount and assesses the institutions, and especially the organized groups providing venture capital to new high-technology enterprises, that have evolved to embrace those risks. It notes that the pace of economic growth achieved during the past two centuries has been accompanied by a high and fairly constant rate of growth of scientific and engineering effort. It asks whether such scientific and engineering effort growth rates can be sustained in the future, reaching a pessimistic conclusion with respect to highly industrialized nations but identifying eastern Europe and Asia, with their vast populations and emerging but still unsteady economic growth, as a crucial source of potential future growth.

In sustaining economic growth during the next century, business enterprises and governments will play vital roles. The United States has led the world in developing venture capital provision organizations that have fueled multiple technological revolutions. Canada and the United Kingdom are cultivating similar institutions, but continental European nations still have much to learn and accomplish in that domain. Because individual enterprises find it hard to appropriate the benefits from fundamental scientific advances, industry is unlikely on its own to provide sufficient financial support for much-needed basic scientific research. Governments will have to fill the gap. In the industrialized nations, maintaining and enhancing the quality of educational institutions, from the preschool stage through university, will be crucial to renewing and revitalizing scientific and engineering work forces. To maintain incentives inducing able students to specialize in scientific and engineering disciplines, business firms may have to reconsider the structure of their compensation systems. Even so, for nations such as Canada, the United Kingdom, and the United States, immigration policies that are particularly open to highly qualified scientists and engineers can be an important complement to domestic educational policies. And if the potential for economic growth benefiting the much larger number of individuals living in less prosperous nations is to be realized, the leading industrialized nations must provide both encouragement and cooperation.

The British–North American Committee believes that these issues merit further study and development by both business enterprises and governments if technology-based economic growth is to be sustained in the coming century.

The BNAC members thank the National Policy Association and in particular Dr. Richard Belous of that organization for their support and encouragement to Professor Scherer in the preparation of this book.

SIR MICHAEL BETT
JOHN MCNEIL
ROBERT ROGERS
Joint Chairs of the British–North American Committee

Contents

New Perspectives on Economic Growth and Technological Innovation

Introduction 1

THE APPROACH OF a new millennium affords an opportunity to contemplate long-term trends. No phenomenon in economic history is more striking than the difference between the past three centuries and the several thousand years that preceded them. Throughout recorded human history, scientific and technological knowledge advanced by fits and starts. But in Europe, roughly three centuries ago, a marked acceleration in the application of science and technology to agriculture, industry, transportation, and other fields of economic endeavor began. This trend was accompanied by dramatic changes in standards of material well-being. The turning point in these relationships is commonly associated with the first industrial revolution, though there were antecedents.[1]

The long-run consequences of those changes can be assessed by attempting to measure, however imperfectly, growth in the quantity of goods and services available to the average citizen—that is, growth in what economists call real income per capita.[2] For the period between 1000 and 1700 A.D., David Landes estimates, real income per capita in western Europe roughly trebled. This trebling over 700 years implies an average 0.16 percent rate of annual increase. Between 1700 and 1750, the apparent growth rate rose to 0.4 percent. Then, during the next century and a half, it escalated to somewhere in the range of 1.2 to 1.5 percent a year. Whereas it had taken more than four centuries for the real income of an average citizen to double during the Middle Ages, the doubling period with a steady 1.4 percent annual growth rate was reduced to fifty years. As a consequence, the average British, Canadian, American, French, or Japanese citizen today enjoys a panoply of goods and services that would evoke wonder and envy from sixteenth-century European nobles.

As these transformations became evident, scholars sought with varying degrees of success to understand why they had happened and what principles underlay them. And now, as the millennium draws near, the question burns, Can continuing growth be sustained, and if so, at what rates and for how long?

Adding interest to such questions, historical and prospective, is the fact that diverse nations have had widely differing economic growth experiences. Figure 1-1 supplies an introductory perspective, showing how real gross domestic product per capita grew (and sometimes fell) during the twentieth century for seven prominent but representative nations—the United Kingdom, the United States, Canada, France, Japan, Argentina, and Korea. The vertical scale, measuring GDP per capita in constant 1990 purchasing parity dollars, is logarithmic, which means that growth at a constant rate is shown by a straight-line trend. The steeper the trend line, the higher the growth rate. At the turn of the century, the United Kingdom, leader of the first industrial revolution, had the highest real GDP per capita among major nations. The United Kingdom was overtaken by the United States for the first time in 1906, and then, after occasional reversals, the United States secured a persistent

Figure 1-1. *GDP per Capita, Seven Nations, 1900–94*

Thousand U.S. dollars, 1990 purchasing power parity

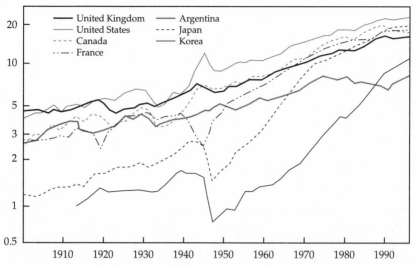

Source: Angus Maddison, *Monitoring the World Economy: 1880–1992* (Paris: Organization for Economic Cooperation and Development, 1995), appendix D.

lead. Japan's GDP per capita was one-fourth that of the United Kingdom in 1900. By growing faster than the United Kingdom and the United States, Japan closed some of the gap in the ensuing forty years and then, following the disastrous consequences of World War II, advanced extraordinarily rapidly after the war to move into near parity with the leading nations. Korea started even farther behind, but after World War II sustained even more rapid and persistent growth than Japan. Argentina began the century as one of the world's most prosperous nations but experienced relatively slow growth, falling back by 1994 to a GDP per capita level only half that of the United Kingdom.

From the tangle of graphic trajectories, table 1-1 extracts in numerical form estimates of the average real GDP per capita growth rates achieved by the seven nations over the full ninety-four years covered by figure 1-1 and also for two shorter intervals—the

Table 1-1. *Average Annual Growth Rates of Real GDP,*
Selected Countries, 1900–94

Country	Percent		
	1900–94	*1950–94*	*1972–94*
United Kingdom	1.35	1.98	1.73
United States	1.82	1.95	1.61
Canada	2.02	2.18	1.63
France	1.96	2.81	1.70
Argentina	1.18	1.18	0.42
Japan	3.03	5.33	2.87
South Korea	3.01	5.80	6.93

Source: See figure 1-1.

forty-four years from 1950 to 1994, and the twenty-two years from 1972 to 1994.[3]

Growth rates during the twentieth century have for the most part exceeded those experienced during the first industrial revolution. A further acceleration of growth in the period after World War II is evident. Since the early 1970s, however, growth has been slower (except for South Korea), triggering widespread concern over the possibility of continued slow growth or even retardation in coming decades.[4] The extraordinarily rapid growth rates of Korea and Japan and the persistently weak performance of Argentina stand out.

These are for the most part success stories, but there are also failures to tell about. Many nations, especially those in Africa, South America, and parts of Asia, have advanced only slowly and as a result have fallen ever farther behind the industrialized nations.

Since the onset of the first industrial revolution, economists have struggled to develop systematic explanations for the causes of growth and to understand the reasons why growth proceeds slowly at some times and in some nations, but rapidly in others. Widely varying theories have been propounded. During the past two decades, a "new growth theory" has taken the economics profession by storm.[5] This book traces the evolution of economic thinking on the economic growth question and examines with particular care the roots of the "new" growth theories. Special emphasis is devoted to technological change, which occupies a key role in some theories of growth but is virtually absent in others.

Traditional Views of Economic Growth 2

H OW DID GROWTH achieve its prominent place on the agenda of economic discourse? This question must be answered in order to understand the new perspectives on economic growth and the role technological innovation plays.

Serious consideration of how economic growth occurs began with the so-called mercantilist economists such as England's William Petty and John Locke and France's Jean-Baptiste Colbert. They saw money, and especially hard specie such as gold and silver, as a source of national wealth, to be accumulated assiduously. To build gold and silver reserves, they argued, it was desirable to achieve a positive balance of trade through aggressive export promotion and tariff or quota restrictions on imports. The specie obtained in this way would reduce interest rates and spur investment in the home market, leading to an increase in domestic employment (implicitly assumed otherwise to be deficient) and hence enhanced prosperity. A positive trade balance was also believed to raise employment directly, thus killing two desirable birds with one stone.[1]

Adam Smith

Adam Smith challenged mercantilist logic and argued persuasively for free-trade policies. He was also the first economist of lasting significance to emphasize raising the well-being of consumers as the principal goal of sound economic policy: "It is the great multiplication of the productions of all the different arts . . . which occasions, in a well-governed society, that universal opulence which extends itself to the lowest ranks of the people."[2]

Smith's was an optimistic message. He saw the prospects for "universal opulence" as essentially unbounded, if markets were freed to guide the allocation of resources and reward producers who, pursuing their own profit-oriented self interest, satisfied consumers' wants. If governments confined themselves to providing national defense, maintaining order, administering justice, and educating the populace and refrained from placing restraints on commerce, both internally and internationally, economic growth would occur naturally as a consequence of three main phenomena.

Smith stressed the importance of what he called "the division of labor," that is, the increase in productive capabilities that follows when each gainfully employed individual specializes in a relatively narrow set of activities, attaining proficiency and minimizing the amount of time spent shifting from one task to a quite different one. To illustrate his case, he cited the productivity gains that resulted from the intricate division of labor observed in a pin factory, permitting output per worker 200 or more times greater than would have been possible if one worker carried out all the diverse operations of pin-making. These productivity gains increased, Smith wrote, with the extent of the market served: the larger the market, the more finely tasks could be subdivided, and hence the greater output per worker would be. Greater output per worker meant more affluence and hence more demand, increasing even more the size of the market and hence the possibilities for division of labor in a kind of virtuous spiral. In addition, free trade opened up markets of international scope, augmenting further opportunities for the specialization of functions.

Writing at a seminal stage of what we now call the first industrial revolution, Smith was keenly aware that improvements in technology played a key role in raising workers' productivity. In Smith's view, technological progress was in turn accelerated by the division of labor: "The invention of all those machines by which labour is so much facilitated and abridged, seems to have been originally owing to the division of labour. Men are much more likely to discover easier and readier methods of attaining any object, when the whole attention of their minds is directed towards that single object, than when it is dissipated among a great variety of things."[3]

But pursuing his metaphor further, Smith foresaw the emergence of present-day research and development laboratories:

> All the improvements in machinery, however, have by no means been the inventions of those who had occasion to use the machines. Many improvements have been made by the ingenuity of the makers of the machines, when to make them became the business of a peculiar trade; and some by that of those who are called philosophers or men of speculation, whose trade it is not to do any thing, but to observe every thing; and who, upon that account, are often capable of combining together the powers of the most distant and dissimilar objects. In the progress of society, philosophy or speculation becomes, like every other employment, the principal or sole trade and occupation of a particular class of citizens ... and the quantity of science is considerably increased by it.[4]

To put those labor-enhancing machines in place, Smith recognized, required a third contribution: investment, or the accumulation of capital, which was in turn the proclivity of businessmen in a profit-oriented economy. "Every increase or diminution of capital, therefore, naturally tends to increase or diminish the real quantity of industry, the number of productive hands, and consequently the exchangeable value of the annual produce of the land and labour of the country, the real wealth and revenue of all its inhabitants. . . . Capitals are increased by parsimony, and diminished by prodigality and misconduct."[5]

Despite his optimism about the prospects for rising material prosperity, Smith could scarcely have dreamed what dramatic results the phenomena he analyzed would have over an extended period of time. In one of the pin factories Smith inspected, ten workers produced 48,000 pins a day, or 4,800 pins per worker. Visiting an English pin factory two centuries later, Clifford Pratten of the University of Cambridge found the average output per worker to be 800,000 pins a day—a productivity increase of 167 times relative to Smithian times![6] A bit of computation reveals the compounded average growth rate of pin output per worker over 200 years to have been 2.6 percent a year—a productivity growth experience not atypical of manufacturing industry more generally.

Ricardian and Malthusian Pessimism

Throughout most of the nineteenth century, however, economists' views of the opportunities for steadily rising economic growth were more pessimistic. Their pessimism stemmed from the writings of David Ricardo and Thomas Robert Malthus.[7]

Among Ricardo's many contributions was a clear statement of what is now called the law of diminishing marginal returns. In agriculture, as more labor worked on a given amount of land, output increased, but at a diminishing rate, as illustrated for the early nineteenth-century English economy as a whole in the top panel of figure 2-1. The total amount of output achieved with varying quantities of workers (counted on the horizontal axis) is measured on the vertical axis by the total output curve (solid line), TQ.[8] The lower panel of figure 2-1 derives from the total output relationship the so-called marginal product curve, ACEM. Although each additional worker adds (up to some high level of employment) incremental output, the increment, or the marginal product, diminishes with additional labor application. Diminishing marginal returns set in as more and more workers are employed for two reasons: because, as the extent of farm cultivation is increased, ever–poorer quality land is drawn into use (at Ricardo's extensive margin), and also because each acre is tended by more workers (more intensively, at Ricardo's intensive margin).

How much labor seeks employment on the land depends in part on how many mouths there are to feed, that is, on the population. Suppose the nation begins with a population such that 5 million workers are employed on the land. Then the marginal product of the 5 millionth worker, read horizontally off the marginal product curve, is approximately 135 bushels of grain. Landlords will pay workers no more than the incremental amount of output they contribute, and since the marginal worker adds 135 bushels of output, the real wage to that worker will approximate 135 bushels a year. Since Ricardo viewed all workers as essentially interchangeable, when 5 million workers are employed in agriculture, each farm worker will receive a wage of 135 bushels a year. With 5 million farm workers employed, the total output (read from the upper panel of figure 2-1) will be approximately 825 million bushels. In the lower panel of the figure, total output is the horizontal sum of all the marginal products of all workers, or the area under the economy-wide marginal product curve between 0 and 5 million workers.[9] Of that total output, 135 bushels times 5 million workers = 675 million bushels, given by rectangular area OBCF, will go to farm workers, who will of course exchange some of their real wages

Figure 2-1. *Equilibrium and Output in the Ricardian Schema*

Employment and total output

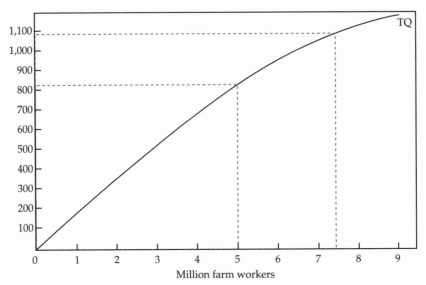

Employment and marginal product

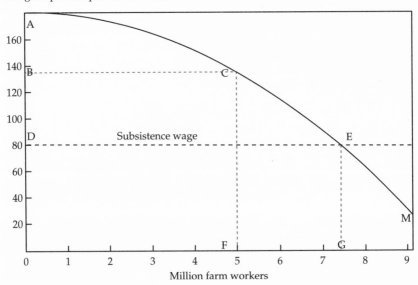

Source: Author's calculations.

(measured in bushels of grain) for manufactured goods. The remaining 150 million bushels—measured by pie slice–shaped area ABC—will accrue as rent to landowners.

Enter now Malthus. Analyzing available population trend data, Malthus concluded that, given adequate nutrition, the population would increase at an essentially geometric rate, doubling and redoubling as the years advanced. Indeed, from his understanding of the law of diminishing marginal returns, he believed that the population could increase more rapidly than the land's capacity to provide food. Thus, under Malthus's view of the world, the 135-bushel wage, with 5 million farm workers employed—well above the 80-bushel "subsistence wage" required for farm families to keep body and soul together and reproduce themselves—would support and encourage population growth. As the population grew, so also would the number of workers employed on farms. Indeed, it would continue growing until the number of farm workers increased to 7.4 million, at which point the marginal product of a worker, and hence the real wage, fell to the subsistence level— barely enough for workers to feed their families well enough to have some members survive high infant mortality rates and keep the population at a stable level. And, as Malthus perceived the intrinsic dynamics of population growth, the subsistence wage was so low that most persons would remain mired in abject poverty.

Meanwhile, as described by the Ricardian-Malthusian schema, landlords prospered ever more while workers were pressed toward the subsistence wage. This too can be seen in the lower panel of figure 2-1. With 7.4 million workers employed, the real wages of all farm workers are given by rectangular area ODEG, of width 7.4 million and height 80 (the subsistence wage), or 592 million bushels. The remaining 488 million bushels, measured by area ADE, go to landowners as rent. This amount is much larger than the 150 million–bushel rent ABC when only 5 million workers are employed. The more employment was extended into ranges of diminished marginal returns, the worse workers fared and the more landlords prospered.

Ricardo and Malthus recognized two ways of escaping these miserable and unfair outcomes. What forced a growing population ever more deeply into the stage of diminishing marginal returns was the ultimate fixity of tillable land. If new land could be made available for cultivation—in sparsely populated North America and Australia—the downward pressure of population on marginal products and hence wages could be postponed. This is illustrated in figure 2-2. With the British Isles alone as the source of land, the economy-wide marginal product curve is M_0, and, with 7.4 million farm workers, the real wage gravitates to the subsistence level of 80 bushels a year. But if new frontiers can be opened up, the marginal product curve shifts to M_N. More total output can be produced by 7.4 million farm workers, and the output of the marginal farm worker rises to 125 bushels, bringing the real wage up to that higher level.[10] The new lands can be tapped through colonial expansion or, as Ricardo argued successfully in seeking the repeal of Great Britain's Corn Laws, by eliminating barriers to the importation of food products from abroad and letting free trade ensure that grain is produced where land is abundant and farmers' marginal labor product is high. British workers gained higher real wages, and, because the workers were fed more efficiently and labor was released from the farms to work in the factories, British manufacturers benefited from a more abundant labor supply and expanded markets for their output. But in the very long run, Malthus warned, the well-fed population would continue growing to force even the new lands into the stage of severely diminished marginal returns. Since at some point all of the world's finite supply of tillable land would be brought into intensive cultivation, the long-run tendency would again be a decline of wages to the subsistence level, at point R in figure 2-2.[11]

Figure 2-2. *How New Frontiers Alter the Ricardian Equilibrium*

Marginal product per worker = bushels

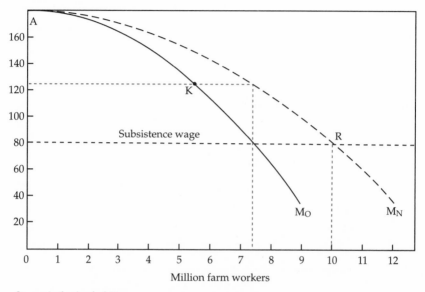

Source: Author's calculations.

The other possible escape route entailed increasing the amount of capital augmenting the efforts of agricultural workers. Using more capital had effects analogous to expanding the supply of arable land, shifting the marginal product curve upward, increasing workers' marginal product, and hence raising equilibrium wages. But eventually the law of diminishing marginal returns holds with respect to increased capital intensity as well as the increased use of labor, and the shifts in labor's marginal product curve with increasing equal-sized "doses" of capital dwindle to the point at which it is no longer profitable to invest in additional capital. Meanwhile, the higher real wages that have come from greater use of capital encourage workers to expand their families. In the very long run, the pressure of growing population on land and capital drives the real wage back to the subsistence level, where, according to the classical economists, the economy settles down into an enduring no-growth, poverty-ridden, "steady state" equilibrium. Small wonder then that the profession of economics (or political economy, as it was called then) was dubbed "the dismal science" in 1850 by Thomas Carlyle.[12]

Why Malthus Was Wrong—or Was He?

From the 20-40 hindsight of the present day, we know that Malthus's pessimistic predictions were wrong, or at least, that they have been proved wrong thus far. A much more densely populated Europe feeds itself very well without the open-market policies advocated by David Ricardo. In the United States, whose food-raising land frontier was largely closed a century ago, the fraction of the work force required to feed a steadily increasing population declined from 72 percent in 1820 to 53 percent in 1870, 27 percent in 1920, 4.2 percent in 1970, and 2.4 percent in 1995.[13] The absolute number of individuals working in the U.S. farm sector declined from a peak of 11.4 million in 1920 to 3.4 million in 1995. Far fewer farmers are producing much more food.

Why was Malthus wrong? The answer involves a complex mixture of increased capital investment, as foreseen by Ricardo and Malthus, and technological progress, which they failed to anticipate. Farmers in all advanced nations of the world work with more machinery of greatly enhanced capability, plant more productive and disease-resistant seeds, strew on their fields more fertilizer in easier-to-apply forms, and (less universally) spray their fields and crops with chemicals to inhibit nutrient-cannibalizing weeds and crop-destroying pests.

A broad picture of what has happened is presented in figure 2-3, drawn from the painstaking research of Yujiro Hayami and Vernon Ruttan.[14] The horizontal axis measures the amount of standardized agricultural output (in wheat metric ton equivalents) per male person employed in agriculture (including farm proprietors). The vertical axis measures the amount of agricultural output produced per hectare of land used in agriculture. The scales are logarithmic, so that a change in the amount of output per unit of input from one to ten covers the same distance as a change from ten to one hundred. Average output/input ratios are presented for five nations over the span of a century from 1880 to 1980 (with hatch marks at ten-year intervals plus one for the war's-end year 1945), during which period all five nations experienced a sharp decline in the fraction of their work force devoted to farming. For each nation, a productivity progress trajectory is shown. The trajectory for the United Kingdom (second from the lower-right) shows, for example, that output per male worker rose from 15.7 wheat metric ton equivalents in 1880 to 116.6 ton equivalents in 1980, while the output per hectare of land rose from 1.1 to 3.0 units. Both labor and land became more productive. The other nations exhibit even larger productivity gains. Japan, for example, moved from 2.3 units of output per male worker in 1880 to 35.0 units in 1980—a fifteen-fold increase in labor productivity. For the United States, labor productivity increased by twenty-two times. The speed of productivity improvement accelerated markedly (shown by longer distances between ten-year markers) after 1930 in all of the nations.

The diagonal dashed lines in figure 2-3 trace trajectories that would have been followed if the progress of productivity had been "neutral," that is, increasing output per unit of land input proportionately with increases in output per unit of labor. In fact, the measured trajectories have a flatter slope, revealing that productivity growth was more labor-saving than land-saving. All of the trajectories slope in a northeasterly direction, however, showing that both land and labor productivity increased. The United States, with the most abundant farmland of the six nations, at all times used production methods that were relatively land-intensive; Japan,

Figure 2-3. *Changes in Agricultural Output/Input Ratios,*
Five Nations, 1880–1980

Output per hectare of agricultural land

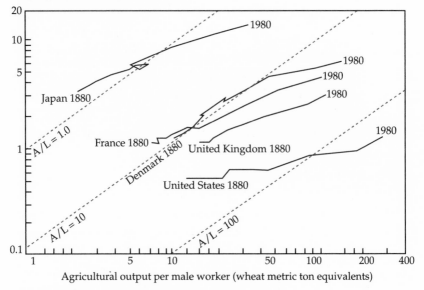

Agricultural output per male worker (wheat metric ton equivalents)

Source: Yujiro Hayami and Vernon W. Ruttan, *Agricultural Development: An International Perspective*, rev. ed. (Johns Hopkins University Press, 1985), appendix tables B-1-5, pp. 467–71. Used by permission of the publisher.

with the least farmland relative to its large population, used relatively labor-intensive methods, that is, with more labor and less land per unit of standardized output. Plainly, progress had differing patterns in the various nations in response to heterogeneous market signals. This is a point to which I will return later.

Figure 2-3 reveals the dynamics of productivity improvement in agriculture over a 100-year span. Figure 2-4 provides a snapshot as of 1980 for thirty-nine nations, a much larger set. The axes are scaled like those of figure 2-3, with output (in wheat metric ton equivalents) per male worker on the horizontal axis and output per hectare of land on the vertical. Again the data show that nations produce their crops in quite different ways in response to market signals. Taiwan, with perhaps the densest population in relation to

Figure 2-4. *Agricultural Output/Input Ratios, Thirty-Nine Nations, 1980*

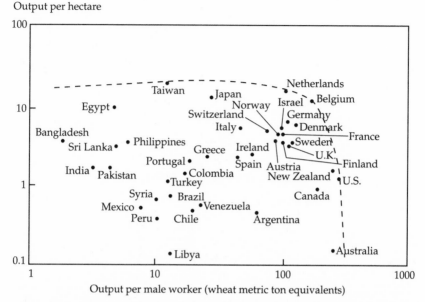

Output per hectare

Output per male worker (wheat metric ton equivalents)

Source: Yujiro Hayami and Vernon W. Ruttan, *Agricultural Development: An International Perspective*, rev. ed. (Johns Hopkins University Press, 1985), table 5-1, p. 120. Used by permission of the publisher.

arable land, uses highly labor-intensive production methods to achieve high output per hectare. Australia, with the most land per inhabitant, uses the most land-intensive methods. Sparsely populated Canada and the United States also use relatively land-intensive methods. The Netherlands and Belgium occupy intermediate positions.

The nations named in the preceding paragraph share a common characteristic. The points depicting their positions on the graph map out a fuzzy boundary (dashed line) of best-practice use, which is called the *production frontier* or *transformation function*. That is, given their choices of whether to adopt relatively land-intensive or labor-intensive farming methods, they go about the task particularly efficiently, using less of the complementary input (labor for the land-intensive nations, land for the labor-intensive nations) than peer nations. Nations with output/input observations to the southwest of the production frontier use more labor *and* more land than peer nations with the same land/labor ratios. They are said to operate inefficiently, the more so, the greater is their distance from the production frontier. Greater distance from the frontier is visibly correlated with what economists call underdevelopment. Underdeveloped nations use much more labor *and* land to produce a standardized unit of agricultural output than highly developed nations. Economic development is all about moving from low output per unit of input to high output per unit of input.[15]

Because the demand for food rises less than proportionately with increases in consumers' income (a phenomenon identified in 1857 by the Prussian statistician Ernst Engel), the rapid improvements in agricultural productivity shown by figure 2-3 have permitted ever fewer farmers to feed the population of the nations they inhabit. The striking changes attributable to rising agricultural productivity in the United States are illustrated by the broken-line data series in figure 2-5. In 1870 roughly half of the U.S. work force was engaged in farming. By 1995 that fraction had fallen to 2.4 percent. A single farm family fed itself and roughly forty other American families, along with originating exports that helped to feed many others overseas. The exodus from farming permitted workers to produce more goods and services in other sectors. In the mining and manufacturing sectors (solid line), which have also experienced above-average productivity growth, similar but more complex patterns materialize. During the late nineteenth and early twentieth centuries, strong productivity growth combined with growing demands for manufactured goods to increase the share of the work force occupied in those sectors.[16] The trend was broken during the 1930s, when the Great Depression curbed the demand for durable goods with special severity. It appeared to resume after World War II, but then reversed, and from 1950 to 1995, the proportion of workers engaged in manufacturing and mining declined steadily from 35.7 percent to 16.3 percent.

Figure 2-5. *Trends in U.S. Sectoral Work Force Shares, 1870–1995*

Percent of total U.S. work force

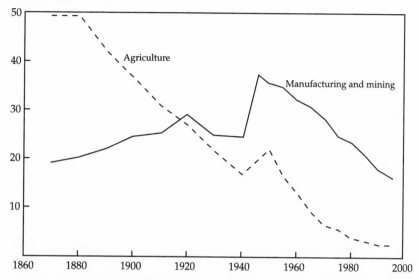

Source: Based on U.S. Bureau of the Census, *Historical Statistics of the United States*, pp. 73–74; and various editions of the *Economic Report of the President*.

The growth of productivity, especially in manufacturing, outstripped the growth of demand for manufactured goods, freeing workers to supply more services, for which productivity improvements have been harder to achieve and for which income elasticities of demand are relatively high.[17]

We return now to the Malthusian theme. Malthus's prediction of convergence to a steady state of poverty-level subsistence incomes has been wrong—thus far. But predictions that the future would unfold in new ways, and that the world was approaching limits to population and income growth, have been made repeatedly by well-informed modern-day Jeremiahs. If diminishing returns in the application of labor and capital to land were not the key barrier to growth, then exhaustion of high-quality hard mineral resources, petroleum, and other fossil energy supplies or even the ability of the earth's atmosphere to absorb the heat and carbon dioxide by-products of energy-intensive production and transportation methods would constrain the opportunities for further growth. Perhaps the best-known of these many pessimistic prognoses was the 1972 report of the so-called Club of Rome, which warned that some time between the years 2000 and 2070, continuing growth of population at the then-current rate of 2.1 percent a year would overwhelm the world's supplies of land and minerals and the capacity of the atmosphere to absorb pollutants.[18] Each time such predictions have been made in the past, they have eventually been proved wrong.[19] Each time, technological progress has bailed humanity out of seemingly impending crises from diminishing returns—for example, through the discovery of new petroleum and gas reservoirs and of ways to extract more petroleum from known reservoirs; improved methods of extracting valuable products from more abundant low-grade deposits of iron, copper, aluminum, and gold ores; techniques for using scarce raw materials more efficiently; the development of more abundant substitute materials; and much else. Furthermore, as nations experienced for the first time the kind of growing prosperity England enjoyed when Ricardo and Malthus wrote, a population explosion followed. But then, as educational levels were raised, citizens' aspiration levels rose apace, and a "demographic transition" to lower population growth rates has tended to occur.[20] In short, we have repeatedly avoided the Malthusian trap. But can we continue to work the magic we have deployed for the past century and a half? On this, doubts cannot be dispelled completely.

Keynesian and Neoclassical Models

Lulled perhaps by the perception that the Malthusian problem had vanished and preoccupied by other concerns such as the instability evidenced during the Great Depression of the 1930s, mainstream economists paid little attention to heroic questions of long-run economic growth during the first several decades of the twentieth century. The early renaissance of mainstream interest bore the earmarks of this changed focus. In apparently independent contributions, Roy Harrod of Oxford and Evsey Domar of the Massachusetts Institute of Technology published new mathematical models of economic growth.[21] Both addressed the problem of how the economy could grow continuously without plunging into recurrent recessions. Each assumed that economic growth depended critically on increasing the stock of capital to keep pace with the growth of the labor force and technological improvements that increased output per labor hour. Assuming a closed economy (that is, one unable to borrow capital from abroad), each argued that the capital needed to complement a growing and more productive labor force had to be raised through saving. If growth were to proceed along an equilibrium path, the rate of saving, which determined the net supply of capital, had to be in balance with the growth rate of the demand for capital. Too much saving led to too rapid capital growth, disappointing business firms' expectations and leading to capital investment reductions that plunged the economy into recession. Too little saving led to stifled economic growth. The economy appeared to be balanced on a knife edge, and it was far from clear that stable growth trajectories could be sustained.

A Ricardian solution to this dilemma came in an article by Robert M. Solow.[22] (On the basis of this and another article, Solow was awarded the Nobel Prize in Economics in 1987.) When capital investment occurred at a rate too high to maintain balance with the growth of steady-state demand, the ratio of capital to labor would rise, diminishing returns would reduce the yield on investments, and firms would respond by curbing their investment to the required steady-state rate. If too little investment occurred, the rate of return on investment would rise, inducing a correction. In this way, long-run steady-state growth could be sustained—if the complications assumed away in Solow's elegant mathematical model did not intrude. Solow's contribution inspired at first a trickle and later a flurry of mathematical growth model-building.

Perhaps more influential at first was Solow's other Nobel Prize article.[23] At the time, most economists believed (and some still do today) that increased output per hour of labor input—that is, the productivity growth that permits rising standards of living—occurs mainly through the accumulation of capital, which lets each worker cooperate with a larger and larger stock of capital as time progresses. Solow set out to test this proposition. He gathered data on year-to-year changes in aggregate output per hour of labor input in the United States over the period 1909–49. Using the mathematical techniques of neoclassical production function theory, he found a way to decompose the growth of output per labor hour into two distinct components, one associated with increased capital use per hour of labor and another with the residual shift unexplained by increases in capital intensity. Although no one should have been surprised,[24] most economists were surprised by his finding that only 12.5 percent (later corrected to 19 percent) of the long-run change in labor productivity could be attributed to increased capital intensity, equivalent to adopting more land-intensive methods of agricultural production along the production frontier of figure 2-4. The remaining productivity growth was impounded in Solow's shift function, which Solow named "technical change." Solow's technical change was a residual component not explicable by measured increases in capital intensity. It could in principle encompass a multitude of causes. However, it was recognized that improvements in technology must have played a major role.[25] Solow's paper was a shot heard 'round the world that transformed the study of technological change into something more than an obscure sideshow in the pantheon of economic research.

The Transition to New Paradigms | 3

THE SEEDS SOWED by Solow and other neoclassical growth theorists fell onto ground that had been prepared by earlier generations of economists whose work was widely viewed as outside the mainstream of economic thought. Among these, Karl Marx and Joseph A. Schumpeter were the most prominent.

Thunder on the Left

Marx is best known for being a revolutionary communist and for extending, without lasting success, Ricardo's labor theory of value. But Marx, unlike other economists writing during the middle decades of the nineteenth century, perceived that the essential genius of capitalism was its ability to combine the accumulation of capital (hence the title of his magnum opus, *Das Kapital*) with an incessant stream of technological innovations. Marx wrote in 1848 that capitalists (pejoratively called "the bourgeoisie") "cannot exist without constantly revolutionizing the instruments of production. . . ."[1] "The bourgeoisie," he continued:

> during its rule of scarce one hundred years, has created more massive and more colossal productive forces than have all preceding generations together. Subjection of nature's forces to man, machinery, application of chemistry to industry and agriculture, steam-navigation, railways, electric telegraphs, clearing of whole continents for cultivation, canalization of rivers, whole populations conjured out of the ground—what earlier century had even a presentiment that such productive forces slumbered in the lap of social labor?[2]

In the basic Marxian schema, capitalists invested in a constant quest for profit. But investment booms led to wage increases and product market gluts, precipitating crises in which the capitalists' profits plummeted. To restore and augment their profits, the capitalists developed and introduced on a massive scale new labor-saving technologies and cultivated new product and territorial markets. The labor-saving investments led to a "reserve army of the unemployed," whose competition with workers still holding jobs kept wages down. But the cycles of rising investment followed by crisis followed by further investment were seen by Marx as growing ever more violent, leading ultimately to revolution by the poverty-stricken workers and a dictatorship of the proletariat. Marx erred in his expectation that cyclical but rising unemployment would prevent workers from sharing through higher real incomes the increased productive potential flowing from technological innovation and the accumulation of capital. But except in that fundamental error, he presented a more accurate picture of the dynamics of the capitalist system than his nineteenth-century counterparts.

Schumpeter

The heir to Marx's view of capitalist dynamics was Joseph A. Schumpeter, an Austrian economist who spent the last eighteen years of his professional life (between 1932 and 1950) as a distinguished member of the Harvard economics faculty. As a twenty-eight-year-old, Schumpeter completed an influential book, *The Theory of Economic Development*, published at first in German (1911) and later in English (1934) and Japanese (1937), among other languages.[3] The book advanced two main themes. First, innovation—including the introduction of new products and production methods, the opening of new markets, the development of new supply sources, and the creation of new industrial organization forms—lay at the heart of economic development, facilitating the growth of material prosperity. Second, innovations did not just happen, but required acts of entrepreneurship—heroic efforts to break out of static economic routines. As he insisted:

> Economic leadership in particular must hence be distinguished from "invention." As long as they are not carried into practice, inventions are economically irrelevant. And to carry any improvement into effect is a task entirely different from the inventing of it, and a task, moreover, requiring entirely different kinds of aptitudes. Although entrepreneurs of course may be inventors just as they may be capitalists, they are inventors not by nature of their function but by coincidence and vice versa. . . . It is, therefore, not advisable, and it may be downright misleading, to stress the element of invention as much as many writers do. [4]

Successful innovations displaced inferior technologies (a process Schumpeter later called "the process of creative destruction") and through imitation and diffusion spread throughout the economic system.

Schumpeter returned to this most fundamental of his ideas in two later books. In 1939, in *Business Cycles,* he argued from an analysis of historical evidence that technological innovations tended to cluster in waves. The three longest upturn phases of these waves, called Schumpeter-Kondratief cycles, coincided roughly with (1) the first industrial revolution, (2) the spread of railroads and ancillary technologies, and (3) the advent of electrical illumination, modern industrial chemistry, and (with delayed effects) the automobile beginning in the last decade of the nineteenth century. In 1942 a new book aimed toward lay readers advanced several additional themes.[5] Two are of interest here. With the improved data emerging from attempts to quantify gross national product, Schumpeter was able to estimate that between 1870 and 1930, real disposable income available for consumption in the United States grew at a rate of roughly two percent a year. If this rate of increase continued for another half century, he extrapolated, it "would do away with anything that according to our present standards could be called poverty, even in the lowest strata of the population."[6] Second, Schumpeter changed his views about the principal sources of innovation. In *The Theory of Economic Development*, he supposed that innovations typically came from small and often new firms operating outside the "circular flow" of existing economic activity. But in 1942 he singled out large and often monopolistic enterprises as the principal engines of technological progress, pressed on the one hand by the forces of creative destruction—those who failed to innovate would be relegated to the sidelines—and possessing on the other hand the superior resources needed to carry out complex technological advances.[7]

Schumpeter attracted a small coterie of disciples. But his arguments were presented for the most part in prose, not in the mathematics that by the 1930s was becoming the leading method of theoretical discourse within the economics profession.[8] In part for that reason, they had little influence on the mainstream of economic analysis, which was preoccupied with applying new mathematical techniques to questions of static resource allocation and to the pressing problems (addressed most notably by Keynes) of recession and the business cycle. Robert Solow's characterization is perhaps not atypical, despite the disclaimer: "Schumpeter is a sort of patron saint in this field. I might be alone in thinking that he should be treated like a patron saint: paraded around one day each year and more or less ignored the rest of the time."[9]

Challenges for New Economic Growth Theories

For those who sought to conceive better mathematical theories of economic growth extending or modifying Solow's contributions, several challenges had to be met. For one, Solow, Harrod, and Domar had assumed that technical change was exogenous—not directly influenced by the economic system, but rather influencing it. It was not unlike manna, raining down from heaven to permit economic growth. This was at odds with Schumpeter's schema, under which entrepreneurs deliberately sought profits by innovating. But more important, even the most casual observation revealed that it was at odds with reality. By 1957, when Solow's second pathbreaking article was published, the U.S. National Science Foundation was collecting annual data on industrial research and development expenditures, which had increased to $7.7 billion, or 1.7 percent of gross national product.[10] Those resource commitments were obviously made with the expectation that technological changes would follow. Consider, for example, this poem written by Kenneth Boulding during a 1962 conference on the economics of industrial research and development:

> In modern industry, research
> Has come to be a kind of church
> Where rubber-aproned acolytes
> Perform their scientific rites,
> And firms spend funds they do not hafter
> In hope of benefits hereafter.[11]

Furthermore, economic research by Jacob Schmookler at the University of Minnesota revealed that the rate and direction of industrial inventive activities were significantly influenced by market forces and in particular, by the ebb and flow of market demand.[12] Using the number of patent applications as an index of inventive activity, Schmookler showed that when investment rose in fields such as railroading, construction, and shoe manufacturing, the number of capital goods inventions relating to those fields increased after a plausible lag; when investment fell, the flow of patent applications dropped in turn. Schmookler did not deny that technological break-throughs less clearly attributable to supply and demand forces had an impact. He found inconclusive indications that the most important inventions *led* investment upswings, while a larger number of improvement inventions *followed* changes in investment, drawn along by the pull of demand. His conclusion that demand forces induced inventions was augmented by evidence of strong correlations between the number of capital goods invention patents and capital investment in twenty-two selected industries for 1939 and 1947. The more investment, the more inventions met the capital goods needs of those industries.[13] From these empirical findings Schmookler argued that differences in the ease of making inventions in one field of technology (compared with another field) mattered mainly in determining the industry from which inventions originated. As Adam Smith had asserted nearly two centuries earlier, firms specializing in industries with rich technological opportunities such as electronics and machine-building produced inventions in response to the demands of quite different industries, to which they sold products embodying their inventions.

Also not to be overlooked was a series of studies showing across diverse industries and within individual firms that the growth rate of output per labor hour—productivity growth—was significantly correlated with the intensity of spending on research and development.[14] Thus, not only did business enterprises respond to market forces in seeking technological innovations through their R&D investments, but they succeeded in influencing the course of subsequent productivity growth through them.

Another challenge facing growth model-builders was the fact that, over several centuries of recorded economic history, the rate of economic growth appeared to be rising secularly, not falling. This was clearly inconsistent with the hypothesis that diminishing returns would prevail as more and more capital was used with given stocks of labor.

The experience of the world's less developed nations posed still another factual challenge. If technological change really did fall like manna from the heavens, it should be an easy matter for the less developed nations to remedy their disadvantage by scooping up manna from the ground or, to alter the metaphor, to pick the technology off the trees like ripe grapefruit. Thus one might expect a fairly rapid convergence of output-per-worker levels among the nations that for some reason led the parade to higher standards of living and those that lagged. Another line of reasoning supported the same supposition. There was abundant evidence that capital investment per worker was lower in the less developed nations than in the highly industrialized nations. If capital accumulation was primarily responsible for economic growth and if the law of diminishing marginal returns held, the less developed nations had carried their investments less far into the stage of severely diminishing marginal returns, and so the return on an incremental capital investment in a less developed country (LDC) should be higher than it would be in leading industrialized nations. This should draw relatively more capital to the LDC, raising labor productivity and causing a convergence of productivity levels toward those enjoyed in industrialized nations. That LDCs often lacked the savings to support vigorous capital investment should not in principle be a barrier to this phenomenon. Wealthy investors in rich nations seeking the highest-yield investment opportunities would channel their investments toward the LDCs, where the marginal returns had diminished less than in the investors' home markets. Capital imported from Great Britain, for example, was an important contributor to U.S. economic development during the nineteenth century.

Productivity growth was in fact much more rapid in some nations that began with relatively low real income per worker than in the United States, the most technologically advanced nation in the world as the second half of the twentieth century dawned. Figure 3-1 characterizes William J. Baumol's analysis of the convergence process over a longer span of time for fourteen industrialized nations.[15] The lower the level of GDP per work hour in 1870, the faster the subsequent growth rate. Australia and the United Kingdom, with the highest initial productivity and hence nearest the early frontier of technological possibilities, experienced the slowest growth rates.

But the convergence hypothesis applied less well for a large number of other LDCs. Indeed, for many, there was no convergence at all; they remained mired in poverty, with little or no trace of accelerated growth.[16] Many of those low-growth nations are the ones shown far below the production frontier in figure 2-4, with especially low output/input ratios in the production of agricultural goods. An intellectually satisfying economic growth theory had to confront this anomaly among others.

The Emerging New Theories

One significant step taken in developing the so-called new economic growth theories was to recognize that physical capital was not the only kind of capital used in production. "Human capital"—that is, the augmentation of basic human skills through education and training—was seen to be at least as important, and so a human capital variable was added to economy-wide output equations (called aggregate production functions) that previously included as inputs only physical capital, labor, and a shift variable measuring technical change. The insight was not new. It had been offered in a variety of forms by earlier economists, most notably by Chicago's Theodore Schultz, who was awarded the Nobel Prize in Economics for his contributions.[17] Schultz was inspired by, among other

Figure 3-1. *Productivity Growth Rates from 1870 Base Level*

Percentage annual growth rate, 1870–1979

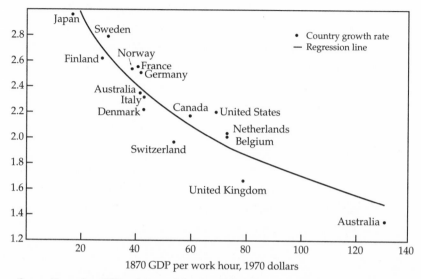

Source: Drawn from William J. Baumol, Sue Anne Batey Blackman, and Edward N. Wolff, *Productivity and American Leadership: The Long View* (MIT Press, 1989), p. 94, using data compiled by Angus Maddison. Used by permission of the publisher.

things, observing the rapid recovery of production in Germany and Japan following World War II despite massive destruction of physical capital. What was not destroyed, Schultz realized, was the capital embodied in managers' experience and workers' accumulated skills, which provided the basis for an industrial renaissance. A subsequent Nobel laureate from the University of Chicago, Robert Lucas, developed an influential new model of economic growth based on the assumption that human capital, unlike physical capital, might be augmented with constant rather than diminishing marginal returns, permitting economic growth to continue indefinitely.[18]

A more dramatic leap was attempted in two papers by Paul Romer.[19] Like others before him, Romer addressed paradoxes posed by Solow's assumption that technical change was essentially exogenous. A common (non-Schumpeterian) interpretation of that assumption had been that technical changes came in the form of scientific discoveries made through academic research, or from government laboratories largely devoid of crass commercial motives. If technical change rested on scientific discoveries, most economists recognized, the peculiar properties of knowledge as an economic good came to the fore. Knowledge is believed to be one of the closest things known to a pure public good, that is, something whose use by one person does not preclude its use by other parties—a property known as nonrivalry—and whose use by other parties is difficult to prevent (except through the maintenance of strict secrecy)—a property known as nonexcludability. These insights were also not new. Indeed, they were achieved in full by the United States' first patent examiner, Thomas Jefferson:

> If nature has made any one thing less susceptible than all others of exclusive property, it is the action of the thinking power called an idea, which an individual may exclusively possess as long as he keeps it to himself; but the moment it is divulged, it forces itself into the possession of every one, and the receiver cannot dispossess himself of it. Its peculiar character, too, is that no one possesses the less, because every other possesses the whole of it. He who receives an idea from me, receives instruction himself without lessening mine; as he who lights his taper at mine, receives light without darkening mine.[20]

To escape the problems of having such a public good as an input in aggregate production functions and the paradoxes posed by the failure of LDCs to take advantage, Romer postulated that technological progress in industry requires concerted, profit-oriented activity that yields two distinct components: specific designs embodied in products that can be patented and produced, excluding rival firms from the same activity; and the knowledge of those designs that is essentially a public good. To create new designs, ordinary labor and capital do not suffice; human capital must be devoted to the task. The human capital is made more productive by interacting with the stock of knowledge, which includes knowledge of all designs previously achieved along with the scientific knowledge published by academic researchers. The more knowledge there is, the more productive R&D efforts using human capital are. The new products based on those efforts are ordinary economic goods. Use of a specific unit by one actor precludes use by another, and, through patent protection, other firms can be excluded from using the designs to manufacture identical competing products. But the design knowledge produced as a by-product of the effort "spills over" into the general pool of knowledge, where it can be used by an indefinite number of other firms in the creation of other new products. Each of those products is different from all others, and each therefore can be produced by multiplying ordinary capital and labor indefinitely in constant proportions without running into diminishing marginal returns.[21] Meanwhile, the pool of nonexcludable design knowledge grows, making further R&D efforts (using excludable human capital) more productive and hence facilitating the creation of still more new products. The more human capital an economy possesses, the more productivity-enhancing new products it can develop and the more the design knowledge stock is enhanced, permitting sustained economic growth in a virtuous spiral. Economies with very little human capital, such as the typical less developed country, are unable to orchestrate a comparable interaction between their (scarce) human capital and the pool of design knowledge, no matter how large the latter becomes, and therefore they are unable (at least internally) to sustain the production of capital goods essential to rapid economic growth. Therefore, underdevelopment and low productivity persist.

Romer's model of economic growth remains lean and elegant, capable of considerable elaboration. Inspired by it and other contributions emphasizing the role of knowledge spillovers in enriching the fertility of scientific research and the opportunities for productivity growth, economists began looking under every lamppost for empirical evidence of knowledge spillover effects. The search has yielded a plenitude of plausible linkages, some close to Romer's characterization, others not.[22]

Spillovers and Unappropriated Benefits from R&D

It was well understood from at least the early 1960s that successful new industrial product development efforts generated benefits accruing to entities other than the firms carrying out the development—that is, in a sense more broadly construed than the Romer paradigm, "spillovers."[23] Figure 3-2 illustrates the basic theory. Suppose a new product is conceived and marketed to fill needs previously unserved. The result is that a demand curve, D, for the product and an incremental cost function, C-MC, come into existence where none existed previously. Suppose the new product is sold under conditions of pure monopoly. A relatively high price, OP_M, is set, leading to the consumption of OX_M units of the new product. The price exceeds incremental production cost by P_MC per unit on OX_M units, so the contribution to the repayment of the innovator's R&D costs plus profit is given by the area P_MBEC. But the area under the demand curve up to the amount of output measures the value of the product, as perceived by buyers. Thus the triangular area ABP_M measures what economists call consumers' surplus—that is, value above and beyond what consumers must actually pay for the product. It follows that some benefits from a new product are not captured by the product's seller.[24] In a sense, they spill over to consumers. If, contrary to the assumption of pure monopoly, some competition develops in the sale of the new product, the price may be competed down, say, to OP_C. Now the profit contribution realized by producers shrinks to P_CFGC, while the consumers' surplus triangle expands to AFP_C. Compared with the monopoly equilibrium, consumers are better off, while producers

Figure 3-2. *Producer and Consumer Benefits from Innovation*

Cost, price per unit

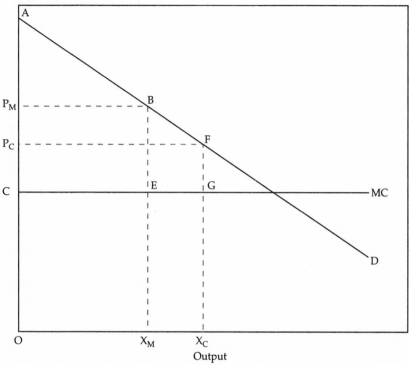

Source: Author's calculations.

are worse off. Economists describe this situation by saying that the producers are appropriating a smaller fraction of the benefits from their new product.

Pioneering research by Edwin Mansfield and colleagues examining seventeen technological innovations by U.S. companies revealed that, on average, nearly half the net benefits from new products and processes were appropriated by innovators, with the rest accruing to consumers and other entities.[25] Transforming the innovators' benefits into rates of return on R&D investment, Mansfield and his colleagues calculated that the median new product or process yielded a 25 percent before-tax return on R&D and related investments. Adding the gains realized by other parties to those of the innovating firms, the authors found that the median *social* return on R&D investment was 56 percent—that is, more than twice the innovator's return.

Pursuing a suggestion by Jacob Schmookler, F. M. Scherer constructed a detailed matrix of how new technology flowed through the U.S. economy from the firms carrying out R&D in 1974 to business firms, government agencies, and consumers using the products stemming from that R&D. Figure 3-3, drawn from that effort, illustrates the broad pattern of interindustry flows.

Most R&D was performed in manufacturing industry.[26] Roughly a fourth of that R&D is so-called process R&D, aimed at improving the performing firm's internal productivity. New technology embodied in intermediate components, raw materials, and (especially) capital goods of nearly as great a magnitude flowed from one part of the manufacturing sector to other manufacturing industries, enhancing the productivity or product quality of the receiving industries. But an even larger amount of new technology—corresponding to half of manufacturing industry's 1974 R&D investments—was embodied in products flowing from the manufacturing sector to other sectors of the economy, working its growth-enhancing magic in those other sectors.

Because, following the logic of figure 3-2, those interindustry product sales transmit surpluses from the R&D–performing industries to industries that purchase new products, statistical analyses of the relationship between R&D spending *within* industries and the productivity growth of those industries tend to underestimate the contribution of technological change to productivity growth. Using the detailed data underlying figure 3-3, Scherer estimated an average 1973–78 rate of return on R&D investment of 70 to 74 percent in the industries *using* the products emanating from 1974 R&D.[27] Much of that high return was attributable to the use by non-

Figure 3-3. *Technology Flows in the U.S. Economy, 1974*

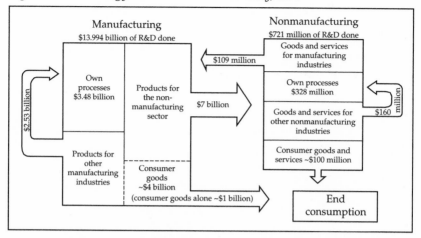

Source: F. M. Scherer, "Interindustry Technology Flows in the United States," *Research Policy*, vol. 11 (August 1982), pp. 227–45, with permission from Elsevier Science © 1982.

manufacturing enterprises of R&D–embodying capital goods developed in the manufacturing sector.

The embodiment in capital goods of technological advances achieved through research and development makes it difficult to disentangle the effects of technological change from the impact of plant and equipment investment in analyzing the sources of economic growth at the economy-wide or broadly defined industry level. Some analyses afflicted by this problem have concluded that most of the economic growth achieved in the most dynamic economies of east Asia was attributable to vigorous growth of capital investment, leaving little or no residual Solowian role for technical advance in "explaining" increased productivity.[28] But when new plant and equipment embody the latest, most productive technology, it is quite wrong to say that technological advance per se plays an unimportant role.[29] One reason why the most highly developed nations grow slowly is that their industries, possessing extensive stocks of durable capital goods from earlier technological vintages, do not implement new technologies as rapidly as progressive developing nations expanding their capital structures at a rapid pace.[30] Nevertheless, as we shall see shortly, absorbing advanced technology by inaugurating state-of-the-art factories requires complementary human capital—a combination not all developing nations have been able to muster.

Interindustry (and international) transfers of productivity-enhancing new technology occur within the context of market transactions and are mediated by the market mechanism. Technology also spills over from originating enterprises to other firms through market-mediated technology and know-how licensing transactions and through the hiring of consultants' services. On the international plane, U.S. resident companies received in 1995 from foreign residents $27.0 billion in royalties and fees (80 percent of them from affiliated overseas branches) and paid out $6.31 billion.[31] Domestically, Dietmar Harhoff has shown, technologically progressive "upstream" firms have incentives to transfer technology gratis to their downstream industrial customers in the expectation that the customers' intermediate product purchases will be increased, raising upstream sales and profits.[32]

Last but not least, countless spillovers occur through the public-good properties of scientific and technological knowledge. Because they take place outside the scope of market transactions, they are what economists call externalities. Companies study the patents and products of rivals and are spurred to improve on them or invent around them. The R&D staffs of business enterprises track scientific publications and interact in a host of formal and informal relationships with academic and government scientists, carrying useful knowledge back to their commercial activities. Academic investigators learn from each other in a web of knowledge exchanges that defy quantification. Such spillovers, which are most like those postulated in Romer's formal model of economic growth, have been the subject of numerous empirical studies.[33] Significant links have been found, either through direct observation or statistical analysis, to the physical or scientific productivity of an industry, firm, or university group from the previous scientific or technological contributions of other groups located near the subject group in geographic space, academic discipline, or product characteristics. It seems clear that a phenomenon of considerable magnitude is being assessed.

To the extent that spillovers occur because firms conducting R&D cannot exclude others from using the resulting increments to the knowledge pool, there exists a possibility of incentive failures, since users do not pay for public goods whose production requires appreciable resource investments. Under the standard assumptions of pure and perfect competition among sellers, revenues from the sale of goods are just sufficient to cover the incremental cost of producing those goods, so costs sunk in creating ancillary public-goods knowledge will not be recovered. The assumption of perfect competition is incompatible with the assumption by Romer and others that economic growth depends on massive industrial knowledge spillovers.

One way out of the dilemma is to have the government conduct the knowledge-generating research itself or to subsidize its conduct by private enterprises. That solution poses well-known incentive problems, although for research of a fundamental nature, subsidization (preponderantly to academic investigators, rather than industry) seems unavoidable.

An alternative escape route was elected by Paul Romer, who assumed that product design activities in industry are conducted under conditions of monopolistic competition, that is, where firms possess sufficient monopoly power in pricing their new products to recover not only direct production costs, but also the costs sunk in developing those products and generating knowledge spillovers.[34] Patent, copyright, and trademark protection may bolster the innovator's monopoly position. This monopoly power is not unbounded, however; it is constrained by the emergence of rivals offering technologically differentiated substitutes and leapfrogging to still better products (Schumpeter's "creative destruction"). Scrapping the assumption of pure and perfect competition, on which much of economists' general equilibrium theory is built, was a step taken reluctantly. But it was necessary if theories of economic growth consistent with continuing long-term increases in material well-being were to be formulated. And it comported much better with evidence on the competitive conditions under which most real-world industries actually operate.[35]

Market Size, International Trade, and the Progress of LDCs

Just as Adam Smith inferred that a richer, productivity-enhancing division of labor was possible in large markets, enhanced market demand and supply generate virtuous growth spirals in models like those postulated by Paul Romer. In a larger market, there is room to develop more differentiated new capital goods and materials while still covering the fixed costs of R&D, and the development of any given product can be carried to higher levels of perfection (that is, farther into the stage of diminishing marginal R&D returns).[36] More R&D means more knowledge spillovers, which enhance the fertility of further R&D. In Romer's formulation, the rate at which new designs and knowledge spillovers emerge depends also on the amount of human capital available. Thus a nation such as Brazil, with 157 million inhabitants but relatively few scientists and engineers, would be expected to make fewer indigenous technological innovations than Japan, with a smaller population (125 million) but far more technically trained employees.

International trade has several important additional effects in this context. For one, when markets are open internationally, firms located in relatively small nations need not be constrained by the limited opportunities within their home markets. If they can plausibly view the world as their market, and if they have sufficient human capital, they can undertake R&D projects of great scope and wide diversity.[37] Second, having to compete with the best offerings from other nations also forces companies to strain for new products of superior quality and reliability—if they are tough enough to withstand the competitive pressures.[38] But competition can cut two ways: those who come up with products that prove to be inferior or too late may be forced to withdraw.[39] Third, competing on a world scale facilitates tapping worldwide knowledge pools, which may encompass more opportunities than local pools. As Gene M. Grossman and Elhanan Helpman observe, exporters may learn ways that the wares they are selling can be used more productively, and importers may show how an exporting firm's products can be improved to satisfy needs in target markets.[40]

International trade has been especially important to the increasing technological proficiency of some developing nations.[41] Their shift into modern industries has been facilitated by imports of technologically advanced machinery from more highly developed nations and, at least equally important, from the technical advice that accompanied the machine purchases. To earn the foreign exchange needed for machinery and technologically advanced intermediate materials purchases, LDCs without abundant natural resource endowments such as Korea and Taiwan had to build viable manufacturing export industries, and vigorous participation as both buyer and seller in world markets accelerated their acquisition of technological know-how. These reciprocal dynamics help explain the rapid economic growth of nations pursuing manufacturing export-led growth strategies. Nations that tried to wall themselves off from imports and build industries capable of meeting their domestic needs autarkically—pursuing so-called import substitution policies—have tended to be much less successful in absorbing the best modern technology and hence in raising local productivity. These strategy differences had much to do with the rapid convergence of real output per capita in a few less developed countries toward the standards of industrialized nations while other nations remained mired in low productivity and slow economic growth.

Contrary to the implicit assumptions of early growth theories, absorbing frontier technology is not easy; advanced technology is not, in fact, a free good. Even when enterprises that command state-of-the-art technologies are willing to transfer them (for instance, when they are embodied in capital goods, or under licensing agreements), it takes concerted effort to receive them effectively. Technology transfer activities are not likely to be very successful when the recipients are passive, expecting modern techniques to be handed over on a silver platter (or through a contract for construction of a turnkey plant). Much more effective are technology transfers from the home laboratories and plants of multinational enterprises to LDC branch plants. There is debate, however, whether best-practice techniques continue to diffuse out to indigenously owned plants.[42] Technology transfer to indigenous LDC enterprises appears to have been most successful when the recipient nation has a strong cadre of technically trained personnel and when those individuals work closely with machine suppliers, know-how licensors, and plant builders, querying transferor representatives about the whys and wherefores of each detailed technical choice. South Korea offers an outstanding example of an LDC that has industrialized rapidly pursuing this strategy.[43] Many other less developed nations have been surprisingly slow in learning and applying the lessons from the Korean experience. There is evidence, too, at least from the diffusion of technology within highly industrialized nations such as the United States, that business firms are better able to absorb advanced technology when they perform at least some R&D, including basic research, internally.[44] Thus interactions between human capital stocks and the pool of knowledge are indeed important.

The Role of Government and Institutions

From an initial emphasis on physical capital, human capital, self-generated technical change, and spillover knowledge in austere economic models of economic growth, scholars have reached out for other variables to explain why some nations have been so much more successful than others at achieving economic growth. Prominent among them are variables attempting to measure how conducive the nations' institutional environment and government are to enterprise and innovation. Scrunching the social and governmental fabric of a nation into quantitative measures is no easy task. The various candidates have included indexes of political instability (proxied by the frequency of revolutions and political assassinations), the black-market premium on foreign exchange (reflecting government-induced price distortions), the ratio of government consumption to GDP; and subjectively assessed rankings of national government propensities on such dimensions as expropriation risk, corruption, the extent to which government is carried out according to rules of law, and the frequency of contract repudiation by governments.[45] Many of these variables exhibit explanatory power in statistical analyses taking real GDP growth rates for a sizable sample of nations as the dependent variable. Typically, the more distortionary and uncertainty-enhancing the governmental environment is, the lower growth rates have tended to be.

Democracy is also said to be conducive to economic growth, although there are counterarguments—for instance, that dictatorships (such as those of Stalin and South Korea's Park Chung Hee) might be more effective in mobilizing resources for growth or persuading business leaders to invest in local development rather than Swiss bank accounts. In an exploratory paper, Ishtiaq Mahmood argued that the relationship between democracy and growth may be nonlinear.[46] An authoritarian regime may facilitate growth (although it may, alternatively, squander resources on wars or profligate consumption). However, as income per capita rises and a nation's static and especially dynamic resource allocation challenges take on increased complexity, authoritarian governments become less and less capable of sustaining further growth. For a sample of 116 nations with growth rates measured over the years 1960–90, and using subjectively estimated indicators of political freedom, Mahmood found weak but statistically significant support for the hypothesized "inverted U" relationship.

Growth over the Very Long Run

The growth models conceived by Paul Romer recognize that there may be diminishing marginal returns in the ability of R&D resources to generate new products and production processes, given the stock of knowledge.[47] However, as the overall scale of the world's R&D efforts rises, the stock of public-goods knowledge also grows, enriching R&D efforts so as to offset tendencies toward diminishing marginal returns and hence permitting permanently positive material prosperity growth.[48] This is by no means certain. If each packet of new public-goods knowledge is applicable to only a limited array of industrial problems, and if the arrays become more specialized over time, the offset might be insufficient. Also, the invisible hand that guides R&D resource allocations could go astray, stimulating excessive duplication among the research and development projects of individual companies and nations.[49]

It is difficult to know which of these opposing tendencies is stronger. A cautionary note is injected by the research of Derek de Solla Price. He compensated for the lack of official R&D spending statistics by compiling a count of the scientific journals founded over the three centuries beginning in 1665. His evidence, reproduced in figure 3-4, leads to the inference that since 1750, scientific

Figure 3-4. *Trends in the Publication of Scientific Journals*

Number of journals

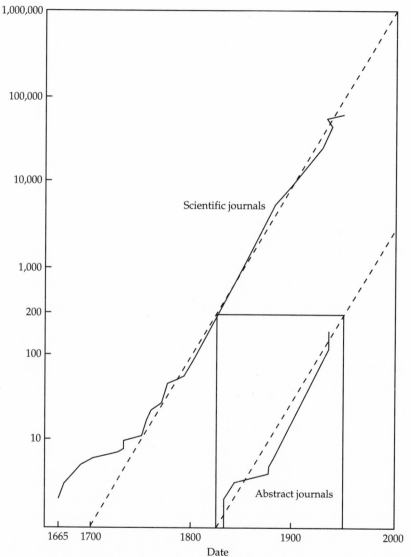

Source: Derek J. de Solla Price, *Little Science, Big Science* (Columbia University Press, 1963), p. 9. Reprinted with permission of the publisher.

effort has been growing at a nearly constant rate of roughly 4.5 percent a year. Cruder data on scientific and technical employment imply similar growth rates. But if scientific activity continues to grow at the rates observed since 1750, Price observes, in less than a century "we should have two scientists for every man, woman, child, and dog in the population." Since this is plainly impossible, Price concludes, "Scientific doomsday is therefore less than a century distant." [50]

During the past two centuries, productivity growth rates in the leading industrialized nations have hovered in the range of one to three percent a year, with higher rates prevailing during the first two or three decades following World War II, after which there was regression to longer-term norms. Is there some law of nature requiring that scientific and technical effort increase at 4 to 5 percent a year in order to maintain annual productivity growth (and hence, with adjustments for work force participation, real GDP per capita) averaging 2 percent?[51] If so, there is trouble on the distant horizon, because the historical R&D growth rates cannot be sustained indefinitely.[52] The most plausible avenue of escape, if indeed one exists, would be to increase the productivity of productivity-enhancing R&D, for example, through better education, more generous allocation of top talent into science and engineering, a richer division of labor in the R&D vineyards, and the enhancement of researchers' productivity through the availability of ever-improving computer capabilities and (most speculatively) changes in human creativity facilitated by advances in the biological sciences.[53]

Interim Evaluation | 4

NOBEL LAUREATE George Stigler remarked on more than one occasion that "it's all in Adam Smith." In a sense, he was right. In a few brilliant pages, Smith anticipated the role creative individuals and specialized research and development organizations would play in propelling technological change and economic growth. Following Smith, however, mainstream economic theory went astray for nearly two centuries, putting far too much emphasis on production relationships from which change, and especially technological change, was largely absent. But still more recently, at first slowly and now on an industrial scale, economists have developed a "new" perspective on economic growth in which new technology not only plays a key role, but responds endogenously to the pull of market demand and the lure of profit.

As always, these new contributions have not persuaded all nonbelievers. Debate continues among other things over how one sorts out what can scarcely be disentangled—the contribution of capital investment per se, as distinguished from the new technology that is typically embodied in new plants and (especially) equipment, whose effective use in turn may require workers with advanced technical knowledge and training.

For managers and policymakers who must confront concrete day-to-day problems, the new economic growth theory has a different shortcoming. It operates at a heroic level of abstraction, assuming aggregated relationships between technical inputs, knowledge outputs, and product outputs that suppress the rich complexity of real-world product development and marketing decisions. It almost never asks, How difficult is it to identify new scientific and technological possibilities with attractive profit prospects? What institutions facilitate the discovery of new technological opportunities; what institutions retard it? How great are the risks? How are they perceived by the individuals who must decide whether to invest time and money in new products and processes? How serious is the appropriability problem—in other words, how large is the wedge separating payoffs to an innovator from total social benefits, including the benefits spilled over from an innovation to consumers and other producers? What strategies can business enterprises pursue to hedge against risks and capture enough rewards from their innovative investments to make the effort worthwhile? What contribution can governments make to ensure that technological progress continues to sustain economic growth? The new growth theories provide inadequate foundations for answers to questions such as these.

At a more aggregate level of analysis, there remain important questions as to how the new theories can be extrapolated to anticipate the likely future progress of industrialized and developing nations. To what extent can the expansion of research and development efforts continue in industrialized nations that are already pushing the frontiers of technological knowledge? What constraints limit the expansion of basic scientific research? Of technological development? To what extent, for example, are development efforts constrained by the orderly progress of scientific knowledge? To what extent are they held back by the availability of creative talent, that is, by the rate at which bright young scientists and engineers enter the workaday world? To what extent do scarcities of human capital limit the convergence of less-developed nations toward the best-practice technological frontier? On what conditions do needed human capital supplies depend? What opportunities for future expansion exist?

Although economic progress will undoubtedly continue even though its causes and constraints are poorly understood by scholars, policies affecting growth processes are likely to be better formulated if the determinants and constraints affecting growth are well recognized. The remainder of this book addresses important facets of these questions.

Investing in Technological Innovation 5

A DVANCES IN TECHNOLOGY, we have seen, are a principal driving force leading to economic growth and increased standards of material well-being. They do not for the most part come like manna from heaven, but must be sought through deliberate activity—activity typically motivated by the anticipation of economic gain. To achieve technological advances, investments must be made in research, development, testing, and dissemination or marketing. The financial investments in turn support another crucial input—skilled creative talent or, in the vernacular of economics, human capital.

The Appropriability Problem

When individuals and private-sector enterprises make investments, they do so in the expectation of eventual profit returns that repay with interest the amounts committed. Here we encounter the difficulty emphasized by those who have worked at a more heroic aggregate level on theories of economic growth. Technological advances generate spillovers—benefits that accrue to entities other than the one making investments in the required research and development. Thus the investors do not appropriate all of the benefits from their investments; others capture a portion. This leads to a potential market failure. Investors may be willing to invest in an innovative project if it yields a best-guess return of at least 20 percent. There are likely to be projects that yield 30 percent, if the returns to all actors in the economy are added up, but for which the returns realized by those who commit the investment capital amount to only 15 percent. If so, the investment will not be made, even though, from the perspective of the economy as a whole, it ought to be made. Most economists believe that the characteristic shortfall of private benefits relative to society-wide benefits leads to systematic underinvestment in advancing technology.

This so-called "appropriability problem" is believed to be most serious for fundamental or basic research and least serious for investments in the development of specific new products or production processes. Knowledge about how the physical world works is often the principal "output" of fundamental research activities. Knowledge, as Thomas Jefferson recognized two centuries ago, comes close to being a pure public good—something whose appropriation by other persons can scarcely be prevented unless strict secrecy is maintained, contrary to all the traditions of science.[1] Most scientific advances cannot be patented. Even when they can be, the protection offered by patents is often so imperfect, or the grounds for ensuring that important new knowledge is not monopolized are so compelling, that widespread diffusion ensues. The histories of two epochal twentieth-century technological advances—the "invention" of the transistor and articulation of how electrons behave in semiconductors, and the "invention" of gene splicing—illustrate this point.

After scientists at Bell Telephone Laboratories discovered the transistor effect in 1947, they secured numerous patents on transistor concepts, various embodiments thereof, and processes for making transistors.[2] Had the Bell System attempted to retain exclusive rights to those inventions, there would have been public outrage, and AT&T, as a regulated public utility, realized that its broader responsibilities mandated widespread dissemination of its findings. Conferences to explain the new principles (but not at first specific production technologies) were convened in 1951 and 1952. Interested parties, domestic and foreign, were invited; at the later conference they paid an admission fee of $25,000, which was treated as an advance against future royalties. Patent licenses were issued to all applicants at royalty rates of 5 percent at most, and often much less. The royalties were reduced to zero in 1956 in the settlement of an antitrust dispute. Thus the Bell System received a tiny fraction of the benefits from the semiconductor revolution it spawned.

Under changes in U.S. law effected during 1980, universities are allowed to seek and retain patents on patentable discoveries made by their faculty and staff, even when the research has been supported by federal government grants. Path-breaking research by Stanley Cohen at Stanford University and Herbert Boyer of the University of California led to the invention of gene-splicing methods, on which three U.S. patents, one granted in 1980, one in 1984, and one in 1988, were obtained.[3] The Technology Licensing Office of Stanford University was responsible for licensing the use of those patents by commercial enterprises. At the end of fiscal year 1994, 290 nonexclusive licenses to the bundle of three patents had been issued. The license terms called for a $10,000 advance payment plus royalty rates ranging from 0.5 percent of sales (on end products such as injectable insulin) to 10 percent on the sale of basic genetic vectors and enzymes. Over the four years 1991–94, the Cohen-Boyer patents yielded to the scientists' universities royalties totaling $75 million. They were the most lucrative of any patents held by U.S. universities at the time. But those "private benefits" constituted only a minute share of the total social value attributable to the Cohen-Boyer patents, which set the stage for a whole new biotechnology industry.

The difficulties of appropriating benefits from more or less fundamental research are so great that private profit-oriented enterprises devote relatively little effort to such research.[4] In 1994, for example, 5.9 percent of U.S. corporations' investments in research and development, broadly defined, were for work classified as "basic."[5] To offset this incentive deficiency, government funding of basic research is considered essential. In 1994 the U.S. federal government provided $16.7 billion for the support of basic research—72 percent of it conducted in universities, 15 percent in government laboratories, and 6 percent in industrial laboratories.[6] Fifty-eight percent of all financial support for basic research in the United States came from the federal government, 25 percent from industry, and 12 percent from state governments (mostly to support work by researchers at state universities) and internal university funds. Altogether, total U.S. basic research expenditures amounted to 0.42 percent of U.S. gross domestic product in 1994. Research and development outlays of all kinds, governmental and nongovernmental, were 2.43 percent of GDP.

While investing in basic research is the forte of government, investing in the development of new products and processes is where industry has comparative advantage. To be sure, government agencies support the development of specific products such as guided missiles, nuclear submarines, and the like to satisfy specialized government agency demands. In 1993 U.S. industry supported with its own funds $85.5 billion of R&D work classified as "development." This can be compared with the $32.9 billion obligated by the federal government for development, two-thirds of which was conducted under contract by industrial firms, and

Table 5-1. *Percentage of Industrial R&D[a] Financed by Government, Selected Leading Nations, Mid-1990s*

Country	Percent of total
United States	16.3
United Kingdom	12.0
Canada	9.5
Japan	1.6
Germany	8.9
France	13.0
Italy	11.8

Source: U.S. National Science Board, *Science & Engineering Indicators: 1998* (Washington, 1998), pp. A-121, A-179.
a. From basic research through development.

95 percent of which came from three agencies, the Department of Defense, the National Aeronautics and Space Administration (NASA), and the Department of Energy. Table 5-1 shows the proportions of all industrial R&D (from basic research through development) financed by government in leading nations during the mid-1990s.

Somewhere between the extremes of basic research and specific new product or process development lie investments in technological advances that have not matured enough to permit commercial embodiment, but that blaze the trail for concrete developments. Investments in such "precompetitive generic enabling" technologies are believed to be susceptible to private-sector market failures nearly as severe as those afflicting basic research. The investment outlays required to bring a technology forward to the point of commercial applicability may be substantial, but once decisive advances have been made, their features are likely to be widely known and appropriable by others, and patent protection may be too weak to deter their use in others' R&D projects. Providing seed-corn financial support for such work was considered by President Bush's scientific advisors to be an important task of government.[7] Similar views were expressed in a white paper on technology policy in the United Kingdom, although the relevant efforts, it was argued, were best supported at the Common Market level rather than by individual nations.[8] In the United States, the Bush initiatives led to the establishment within the National Bureau of Standards and Technology of the Advanced Technology Program, which canvassed American industry for candidate projects in important but insufficiently explored areas of enabling technology. Projects approved after careful screening are supported by a combination of government and company funds.[9] The program grew to a federal spending level of $322 million in 1995, at which point it encountered furious resistance in the 104th Congress. It came close to being terminated, but survived with a budget cut to $182 million in fiscal year 1997. Its longer-term prospects will hinge in part on planned post-project evaluations investigating whether the efforts actually had an appreciable impact on accelerating technological progress.

Patents (supplemented, especially for new developments in computer software, by copyrights) are a policy instrument whose rationale is to offset investors' inability to realize sufficient benefits from their investments in new technology. The patent system is recognized to be an imperfect instrument. Nevertheless, it may be the best solution policymakers can devise to the difficult tradeoff between, on the one hand, maintaining incentives for investment and, on the other hand, fostering the diffusion of new technology's benefits to consumers and to those who might make leapfrogging inventions.[10] Its imperfections include uncertainty as to which of several possibly contending parties will receive patent protection and how much protection patents will afford, the costs of concomitant legal services, and the ease in most cases (but not in pharmaceuticals) of "inventing around" existing patents. The last of these characteristics is so pervasive that, surveys show, most well-established business enterprises consider patents a relatively unimportant means of protecting their innovations from competitive imitation.[11] Much more important in the typical industry are the reputational and learning-curve advantages of being first into the marketplace with a successful new product or process, the possibility of keeping details of the technology secret from competitors, the customer loyalty that comes from providing superior sales and service, and the threat of creative destruction that forces companies to continue innovating or risk being left behind.

The Costs and Risks of Technological Innovation

If investors are risk averse, as most modern capital market theories assume them to be, investments in research, development, and technological innovation may also be discouraged by what are often said to be substantial risks.[12] What are the uncertainties and risks of R&D investments and the strategies available for confronting them?

The Risks

Pioneering research on the uncertainties faced in R&D projects was conducted by Edwin Mansfield and his students at the University of Pennsylvania. In one of the most comprehensive surveys, they obtained information on the outcomes of individual R&D projects in sixteen chemical, pharmaceutical, electronics, and petroleum enterprises.[13] Mansfield and his associates identified three different success probabilities: (1) the probability that a project's technical goals will be achieved; (2) the probability that, given technical success, the resulting product or process will be commercialized; and (3) the probability, given commercialization, that the project yields a return on investment at least as high as the "hurdle rate" that the firm's decisionmakers apply on investment projects more generally. For the sixteen companies combined, the average conditional probabilities were:

Probability of technical success	0.57
Commercialization, given technical success	0.65
Financial success, given commercialization	0.74

Within the technical success category, only four of the sixteen companies had average success ratios below 0.30. The firms' overall average success rate is found by multiplying the three conditional probabilities: $0.57 \times 0.65 \times 0.74 = 0.27$. Thus, on average, 27 percent of the projects initiated ultimately achieved financial success. These, of course, had to provide on average sufficiently high returns to offset the costs of less successful projects.

A somewhat different picture emerges from newer and more detailed research on the distribution of innovation profit outcomes over possible payoff ranges. For ninety-nine new pharmaceutical chemical entities (NCEs) developed in the United States, approved by the Food and Drug Administration, and introduced into the U.S. market during the 1970s, Henry Grabowski and John Vernon estimated "quasi-rents"—the surplus of sales revenues (domestic and foreign) over production, marketing and distribution costs— and discounted them to the date of initial commercial sales.[14] Figure 5-1 arrays their individual product estimates into deciles, ranging from the most profitable drugs in the first decile to the least profitable ones in the tenth decile. The ten most profitable drugs—the "blockbusters"—contributed 55 percent of the quasi-rents from all ninety-nine new drugs. On average, their discounted quasi-rents were 5.6 times the $81 million research, development, testing, and evaluation (RDT&E) cost (including the cost of unsuccessful trials) of the average new drug entity introduced during the 1970s. Drugs in the second profitability decile yielded roughly twice their average RDT&E cost. Drugs in the third profitability decile essentially broke even. For the remaining sixty-nine new chemical entities, all introduced commercially into the U.S. market, average profit returns were less, and often substantially less, than average RDT&E

Figure 5-1. *Distribution of Quasi-Rents Generated by*
New Pharmaceutical Chemical Entities (NCEs) in the U.S. Market, 1970s

Discounted present value per NCE, millions of dollars

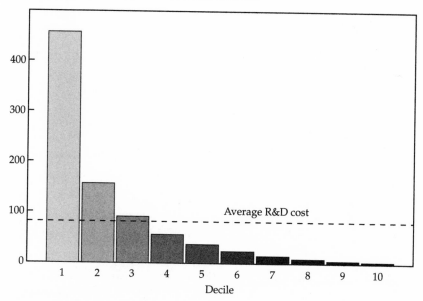

Source: Henry J. Grabowski and John Vernon, "A New Look at the Returns and Risks to Pharmaceutical R&D," *Management Science*, vol. 36 (July 1990), pp. 804–21. Reproduced from original data by permission from Henry Grabowski.

costs. In the aggregate, the "winners" yielded above-cost returns only slightly more than sufficient to compensate for the substantial number of "losers."

That this picture applies more broadly is shown by a survey of 222 particularly valuable U.S. inventions patented during the late 1970s. The inventions passed through a three-fold screen: patents were sought not only in the United States but also in Germany; patents were actually granted in both of those jurisdictions (known to impose relatively high standards for patentability); and annual renewal fees totaling DM 16,075 were paid to keep the German patents in force until their full-term expiration in 1995.[15] Knowledgeable officials in the U.S. companies holding the patents were asked to estimate the minimum price that would have been accept-able in 1980 for conveying to a third party full rights in the patents. Figure 5-2 displays the results. Seventy-six percent of the sample's total value was concentrated in nineteen patents (8.6 percent of the sample by number) with values of more than $100 million. These results, covering a virtually complete spectrum of technologies, are even more extreme than those obtained by Grabowski and Vernon for drugs.

Statisticians call outcome distributions like those observed for drugs and especially valuable patented inventions "skewed," and indeed in these cases, highly skewed. Skewed distributions have the property that the tail essentially wags the dog; that is, relatively few observations account for most of a sample's cumulative real-ized value. A high degree of skewness implies that the law of large numbers works at best poorly, so putting together sizable portfolios of projects is no guarantee that the profit returns of the various pro-jects will average out to a reasonably stable mean. Instead, when the underlying distribution of returns is skewed, considerable vari-ability of portfolio outcomes can be expected.[16]

Another way of seeing how much variability there is in the returns from research and development is to aggregate by company the values of the individual valuable patents discussed above and then to compare those value sums (stemming from only 3.4 percent of the numerous patents received by the typical company in the same time period) with the research and development investments made by the companies in 1976 (in other words, roughly two-and-one-half years before the sample patents were granted). Figure 5-3 (graphed on doubly logarithmic coordinates) provides the result-ing picture. The solid line characterizes the regression equation that best fits the scatter of values of individual firms' patent portfolios. On average, the more firms spent on R&D, the more valuable were their high-value patent portfolios. Indeed, summed patent value rises by roughly 119 percent with each 100 percent increase in R&D

Figure 5-2. *Distribution of the Values of 222 U.S. Patents*

Total estimated value (millions of dollars) of patents in group

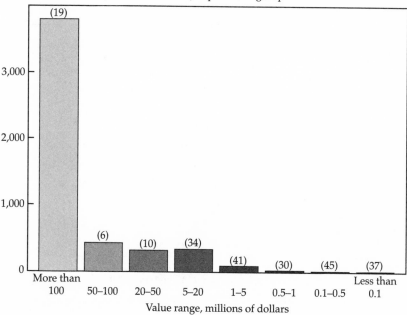

Source: Dietmar Harhoff, F. M. Scherer, and Katrin Vopel, "Exploring the Tail of Patent Value Distributions," Working Paper (Mannheim: Center for European Economic Research, January 1998).

outlays. But there is a wide scatter of outcomes about this central tendency. Some firms (those with marks above the regression line) were overperformers relative to their R&D levels, often by large magnitudes; others vastly underperformed group norms. "Winners" and "losers" are strewn over a wide field.

There are to be sure innovative activities in which the risks, and especially the market risks, are much lower. A considerable fraction of all technological change comes from detailed, small-scale problem-solving on the factory floor—activity unlikely to be called R&D or to yield patents, but whose cumulative impact can be great. Analyzing DuPont's rayon production operations, Samuel Hollander found that a majority share of the impressive productivity advances

Figure 5-3. *Companies' Reported Patent Values Plotted against 1976 R&D Outlays*

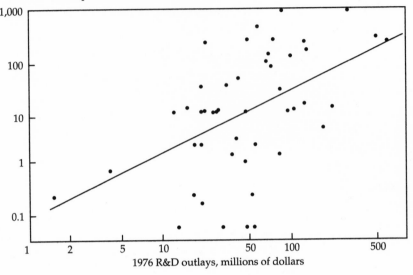

Value of renewed patents, millions of dollars

1976 R&D outlays, millions of dollars

Source: Author's calculations.

achieved between 1929 and 1951 came from "minor" technical changes, most of which stemmed from in-plant technical staff and on which few patents were obtained.[17] Even among the relatively important patents covered by figure 5-2, process inventions were found to yield lower average profits, and their profit outcomes were somewhat less variable, than for product inventions.

Risk-Hedging Strategies

What strategies can organizations pursue to mitigate the effects of uncertainty in research and development? Several possibilities deserve attention.

One standard strategy is hedging bets by maintaining a portfolio of projects. The previous analysis of skewed outcome distributions shows, however, that the portfolio approach is unlikely to come anywhere near securing stable average outcomes.

Second, some risks, and especially the more narrowly technical risks, can be held in check by properly managing the time dynamics of research and development projects. Figure 5-4 illustrates this approach. The pattern of annual spending on a typical relatively ambitious product R&D project tends to follow a slightly asymmetric bell-shaped curve, rising at first and then, after the new product is ready for marketing (for example, at year four), declining. As the project progresses, uncertainty tends to recede. Basic questions such as "Is there a technically feasible product here?" are typically resolved in the early, relatively low-spending stages. Indeed, if favorable answers do not emerge early, the project will be canceled or held in the low-spending phase of exploratory research. As prototypes or pilot plants are built and tested, the spending rate rises rapidly, while uncertainties about how well the prospective end product will perform dissipate. Estimates of what the ultimate product will cost decline roughly linearly with time, but continue (especially when there is learning-by-doing in production) into the early production stages (after year four). If the uncertainties *do not* decline as portrayed here, project cancellation or cutback may be in order before the full burden of the project's R&D costs has accumulated. Timely evaluation and decisionmaking in the phases when spending rises rapidly can save considerable sums. Slowest to decline are uncertainties about market acceptance—for example, how consumers will respond to the product and how competitors will react. Although marketing research can provide some clarification, most of the uncertainties remain until well after the majority of R&D investment, and indeed, appreciable equipment and marketing roll-out investment, has been shouldered. It is the relative lateness of this market uncertainty resolution that leads to the highly skewed profit outcomes observed for FDA–approved drugs and technically successful patented inventions.

Figure 5-4. *Relationship between R&D Project Spending and Changes in Outcome Uncertainty*

Annual rate of R&D spending

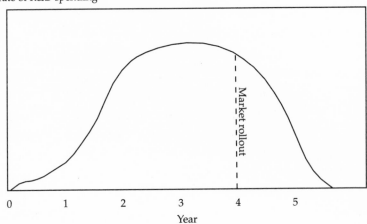

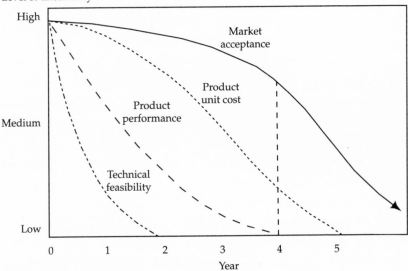

Source: Adapted from M. J. Peck and F. M. Scherer, *The Weapons Acquisition Process: An Economic Analysis* (Harvard Business School Division of Research, 1962), p. 313.

Technical uncertainties and, to a more attenuated degree, market uncertainties, can also be combated by pursuing so-called "parallel path" strategies. In other words, when any particular conceptual or design approach has a low estimated probability of success, multiple approaches are carried to the point at which more confident choices can be made—ideally, before the high spending rates of full-scale final product development are reached. Figure 5-5 provides strategy insights for a highly simplified case (ignoring, among other things, the possible skewness of outcomes) in which each of many possible technical approaches requires spending a $1 million lump on R&D before the success or failure of that approach can be ascertained.[18] Profit-maximizing strategy curves for widely varying initially estimated success probabilities are presented. Two important generalizations emerge. First, the higher the ultimate profitability of a successful outcome is expected to be, the more parallel paths one should pursue. Second, the higher the degree of initial uncertainty is—that is, the lower the estimated probability of success for any individual approach—the more sensitive is the outcome of the parallel paths strategy to choosing the right number of parallel paths.

Individual R&D Project Costs

Whether parallel paths and portfolio strategies can be pursued depends in part on the cost of individual R&D projects. Only sparse systematic evidence exists, and its limitations need to be remembered. The U.S. journal variously called *Industrial Research and Development* or *Research and Development* has for roughly three decades held an annual competition to select the hundred most significant technical innovations of the preceding year. Self-nomination seems to be the norm, and biases no doubt exist. Among other things, the annual lists seldom include major new weapon systems, airliners, or pharmaceutical products—all big-ticket R&D items—

Figure 5-5. *How Choices of Profit-Maximizing Parallel Paths Vary with Uncertainty and Profit Prospects*

Optimal number of parallel experiments

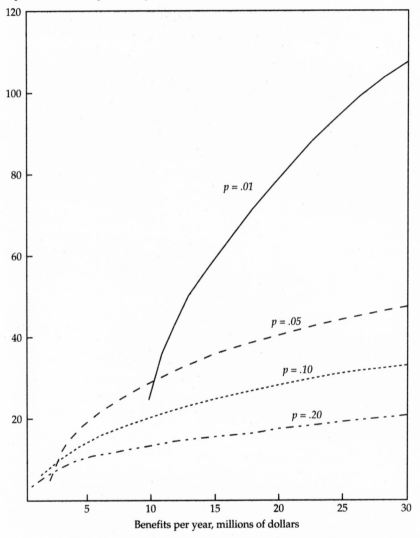

Benefits per year, millions of dollars

Source: Adapted from F. M. Scherer, "Time-Cost Tradeoffs in Uncertain Empirical Research Projects," *Naval Research Logistics*, vol. 13 (March 1966), pp. 71–82. By permission of John Wiley and Sons © 1966.

Figure 5-6. *Average R&D Cost for Winners of Annual "Significant Innovation" Citations from* **Research and Development** *Magazine*

R&D cost, millions of 1985 dollars

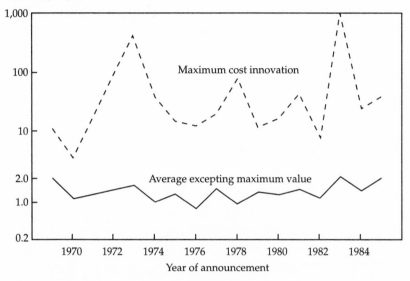

Source: *Research and Development,* annual innovation award issues, 1969–86.

and instrumentation innovations appear with relatively high frequency. Nevertheless, the data presented annually on average and maximum R&D costs of the winning innovations are illuminating. They are summarized in figure 5-6, with dollar values expressed at constant 1985 price levels. What is most striking is that the R&D underlying the average contest winner (excluding from the calculation the single most expensive project) was quite modest, hovering in the range of $1 million to $2 million, with no discernible trend over time. This suggests considerable opportunity for portfolio strategies in companies of moderate size. At the same time, the costs of the most expensive winning innovations were typically high and quite variable from year to year, suggesting that individual R&D project investments, like their profit returns, adhere to a skewed distribution. Most of the project costs cluster at relatively low values, but there is a long tail extending into project costs as high as $1 billion (and if airliner developments were included, multiple billions). For projects in the distribution's "tail," full-blown parallel paths and portfolio strategies are prohibitively expensive. "You bet your company" choices may be unavoidable.

Particularly rich evidence exists on the R&D costs of new chemical entities marketed by pharmaceutical companies. Joseph DiMasi and colleagues compiled detailed data on R&D costs (including outlays for tests) incurred during the clinical trial period on ninety-two new pharmaceutical chemical entities tested in humans between 1970 and 1982.[19] They computed both the out-of-pocket costs for specific drugs and an average cost per success, counting also the costs of failures in the same therapeutic category. The two sets of averages (in millions of dollars per new chemical entity) in four major therapeutic categories and for all tested drugs are shown in table 5-2.

Costs were highest for nonsteroidal anti-inflammatory drugs (NSAIDs), which are taken daily by many patients year after year, necessitating extensive testing for long-term toxicity.[20] High failure rates in testing are implied by the large discrepancy between specific marketed drug costs and the average cost of successes. To achieve a success in the typical case plainly requires a sizable investment. The figures reported, it should be added, exclude the costs of basic research and investigation (through animal tests) of therapeutically interesting molecules in preclinical phases, which in the aggregate are nearly as large as the outlays sustained for clinical testing and development.

Thus the investments required in R&D projects vary widely. Many and probably most projects can be carried out on a modest scale, but some technical challenges compel very large investments undertaken with equanimity only by large companies or risk-loving investors.

Table 5-2. *Average Costs of Successes, Selected Categories of Drugs, 1970–82*

	Millions of U.S. dollars	
Drug category	Cost of successes	Cost averaging in failures
Cardiovascular drugs	16.3	62
Anti-infectives	14.8	50
Neuropharmacological	12.1	61
Nonsteroidal anti-inflammatory drugs (NSAIDs)	22.1	99
All categories	13.7	60

Source: Joseph A. DiMasi and others, "Research and Development Costs for New Drugs by Therapeutic Category," *PharmacoEconomics*, vol. 7 (February 1995), pp. 152–69.

The Sources of Private Funds

Technological innovation is usually risky, and it is often costly. The risks and costs pose problems for securing needed financial support. What economists call "asymmetric information" exists in spades; the companies typically know much more about the prospects and hazards of their projects than outsiders do. And at the early stages of development, the assets that might be offered as security consist mainly of intellectual property—patents, copyrights, trade secrets, and the human capital embodied in R&D staff—property not normally considered good collateral against loans. Institutions have evolved to solve these problems. The solutions have been more successful in some venues—notably, the United States—than in others.

On average, established manufacturing corporations incur R&D costs amounting to roughly 3 percent of their sales revenue. There is considerable variation in this average. In many traditional industries such as food and primary metals, R&D/sales ratios of 0.4 percent are common; in drugs, devoting 16 or 20 percent of sales revenue to R&D is not unusual. For the average well-established corporation, annual expenditures on R&D are sufficiently modest that they can be financed through routine cash flow and, if need be, by resort to outside capital sources willing to provide funds on full faith and credit without detailed inquiry into the specific uses to which the funds will be put. Because alternative funding sources can be tapped, most statistical studies of the linkage between company cash flow and R&D outlays have found little in the way of strong, systematic relationships.[21]

The story is different for smaller and less well-established companies, whose access to capital markets is more imperfect and for whom liquidity constraints often bind tightly. There, careful econometric work by Charles P. Himmelfarb and Bruce C. Petersen reveals, internal cash flow is important. On average, the more cash flow realized by 179 chemical, electrical, instrument, and machinery manufacturers with capitalization in the range of $1 million to $10 million, the more they invested in research and development.[22] Increases in cash flow of 100 percent were followed on average by R&D increases of roughly 67 percent.

For technology-based firms, and especially new or smaller firms, securing bank loans is problematic. Interviews by Bank of England staff members at fifty-nine technology-based U.K. firms during various stages of their life cycle led to the following summary of findings:

> Firstly, very few firms believed that their bank understood the technology involved in their product or the nature of its market. Secondly, firms' perceptions of the standard of bank service varied across different branches of the same banks, rather than between the different banks themselves. Thirdly . . . relationships appeared to improve as firms gained a track record, with the lack of security becoming less relevant or overridden by the subsequent performance of the firm. However, firms noting this improvement said that banks still did not understand their product despite having provided finance for many years.[23]

From this and other evidence gathered through its study, the Bank staff concluded, "Banks are not normally an appropriate source of risk capital for small technology-based firms at their early stages."[24] Germany was found to have even more serious problems in financing small high-technology enterprises, since equity markets are relatively underdeveloped and the bank debt on which the German system of business finance relies heavily is not well adapted to the needs of technology-based firms.[25]

The most highly developed institutions for providing equity capital to new technology-based enterprises are found in the United States. Hundreds of high-technology venture funds review many thousands of equity financing proposals each year, providing to startup enterprises seed finance, experienced advice, and access to managers whose business skills complement the skills of technical entrepreneurs. The pioneer in this genre was the American Research and Development Corporation (ARD), founded in Boston during 1946 and presided over for nearly two decades by General Georges F. Doriot, a charismatic Harvard Business School professor who headed U.S. Army research and development efforts during World War II. ARD's early successes inspired widespread emulation, especially on the west coast, where numerous high-technology venture funds out-competed their New England predecessors and fueled the information technology revolutions of Silicon Valley and the biotechnology revolution centered on San Francisco.[26] The two panels of figure 5-7 illustrate the evolution of ARD's success.[27] The top panel shows how the number of companies in which ARD made equity investments grew over time. From a modest start, additions (lower line) and occasional liquidations raised the net count of ARD investment targets (upper line) into the mid-twenties during the 1950s and the mid-forties in the 1960s. Despite some portfolio risk-spreading advantages, investing in more than about forty companies is considered to overstrain portfolio managers'

Figure 5-7. *The Evolution of ARD's Portfolio*

Number of company investments

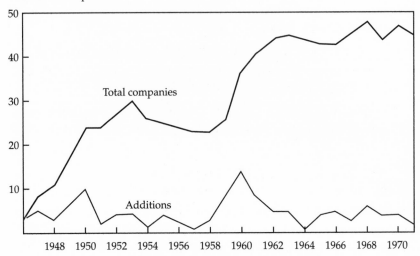

ARD's net asset value per share

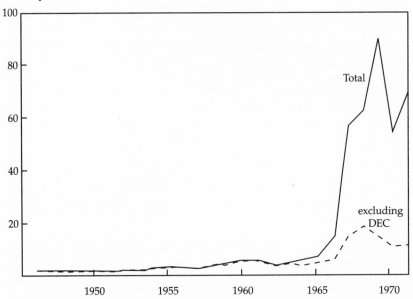

Source: Based on Heidi Willmann, "Innovation in the Venture Capital Industry: A Study of American Research and Development Corporation," term paper, John F. Kennedy School of Government, Harvard University, May 1991. Reproduced with permission.

monitoring ability, so most venture funds (commonly organized as limited partnerships) top out in that range. The bottom panel shows how the combined value of ARD's portfolio investments grew. During the mid-1950s, a few successes (such as High Voltage Engineering Company and defense-oriented Airborne Instruments, Inc.) fueled an appreciable increase in portfolio value. Then, in 1966, the portfolio value exploded. By decomposing ARD's investments into two parts—Digital Equipment Company (DEC) and more than forty other companies—the lower panel reveals that most of the increase was attributable to ARD's 1957 investment of $70,000 in DEC. DEC's great success occurred with the introduction of the first time-sharing computer, the PDP-6, in 1964, and a powerful but inexpensive minicomputer, the PDP-8, in 1965. An initial public offering of DEC's common stock was floated on the New York Stock Exchange in 1966.

The profound effect of a single investment on ARD's overall portfolio value is typical of what happens when the distribution of profit returns from investments in innovation is highly skewed. That this is true more generally is shown by data from a much larger sample of high-technology startup company investments.[28] Figure 5-8 summarizes a study by Horsley Keogh Associates of 670 distinct investments (totaling $496 million) in 460 companies made between 1972 and 1983 by sixteen venture capital partnerships. The terminal portfolio value was calculated as of December 1988, at which time the funds had distributed to their partners $822 million and retained assets valued at $278 million. The thirty-four investments (five percent of the sample total by number) that yielded ten times or more their original value (on average, 19.25 times) contributed 42 percent of the portfolios' total terminal value. Slightly more than half of the investments entailed some loss.

Partly because a relatively few big winners can push industry averages upward, while their absence can mean disappointing overall performance, and partly because investors often respond

Figure 5-8. *Distribution of Gains, 670 Venture Portfolio Investments*

Terminal value of group, millions of dollars

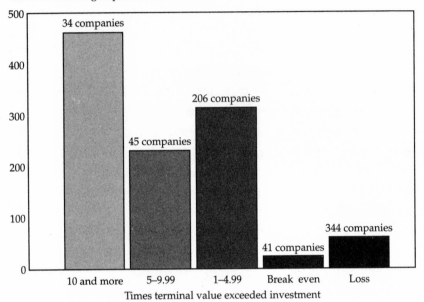

Times terminal value exceeded investment

Source: "Horsley Keogh Venture Study" (San Francisco: 1990).

to evidence of strong past performance rather than unbiased estimates of future performance, there are pronounced cycles in venture investment levels and in the returns from those investments.
When relatively small amounts of capital are available, venture
partners can be highly selective in the investments they make.
When capital is abundant, there is more competition and hence
less time to perform "due diligence," and more decisions fall on less
experienced fund managers.[29] In the United States, high average
returns in the mid-1960s led to a venture investment boom in the
late 1960s, which (along with oil shock recessionary effects) drove
down average realized returns in the mid-1970s, which led to high
returns in the early 1980s, which precipitated a mid-1980s boom
(with annual new capital inflows of $2 billion to $4 billion), followed again by numerous company failures (especially in biotech-

nology) and depressed returns. Boom conditions emerged anew in the mid-1990s, with new capital commitments reaching a record level of $4.2 billion in 1995.[30]

There have been changes over time in the mix of venture fund investors and the orientation of their investments. From the 1950s through the 1970s, most venture fund capital came from relatively wealthy individuals seeking to enrich the yields of their investment portfolios. In 1979, however, the U.S. Department of Labor modified its "prudent investor" rules, permitting pension funds to make limited, well-diversified investments in risky new ventures instead of merely buying blue-chip securities, which had been their traditional practice. A torrent of new capital flowed into venture portfolios, driving down returns and changing the portfolios' qualitative composition.[31] Of the $4.2 billion committed in 1995, $1.66 billion came from pension funds, $784 million from insurance companies, $959 million from endowments and foundations, and $741 million from individuals and families.[32] Some funds also diverted their attention during the 1980s to financing leveraged buyouts, real estate deals, and cable television operations instead of, or in some cases along with, new high-technology ventures.

Typically, high-technology startup companies go through a fairly well-ordered series of financing stages. Their first experimental work and the formation of business plans may be done by the technologist-entrepreneur as a part-time homework adjunct to regular employment. Once full-time efforts commence, they may be financed out of savings or through a mortgage on the principal entrepreneur's home. When additional but still modest injections of capital (in the $100,000–$500,000 range) are needed, they are commonly provided by "angels"—well-heeled individuals who have found a fascinating challenge in high-technology investment.[33] Venture partnerships enter the picture as full-scale prototype and/or clinical testing begins and still larger sums are needed—

typically between $300,000 and $3 million. The number of such investments in the United States during the mid-1990s has been estimated at roughly 3,000 a year. According to Venture Economics, the leading compiler of U.S. venture investment data, 6 percent of venture fund disbursements in 1995 involved seed financing, 17 percent went for startup financing (such as product development and initial marketing), 15 percent for other early stage activities, 42 percent for expansion, 9 percent to finance leveraged buyouts or other acquisitions, and 11 percent for other uses.[34]

If initial technical and market trials yield promising results, the startup company may float an initial public offering (IPO) of its common stock, raising as much as tens of millions of dollars and creating an opportunity for early-stage venture investors to liquidate on favorable terms part or all of their investments. The development through NASDAQ of a nationwide, electronically linked, market for what were previously called "over-the-counter" stocks facilitated such IPOs and led to explosive growth in their numbers.[35]

A company's passing through early technical development and financing stages and launching its first public stock offering does not imply the disappearance of substantial risk. Figure 5-9 tracks the value of hypothetical $1,000 investments made in a sample of ten high-technology IPOs at the time they first "went public" between 1983 and 1986. Cash dividends were assumed to have been reinvested at the time of their payment. The ten are drawn from a larger group of fifty-two companies whose securities continued to be traded into 1995, surviving from an initial sample of 131 IPOs. Five of the ten were deliberately selected as the most successful of the fifty-two; the other five were chosen randomly. The stock value trajectories for the randomly selected five are clustered so tightly in the $0–$2,000 range that they are indistinguishable from one another. In other words, the companies essentially went nowhere. Had one invested in Adobe Systems, Concord Computing, Amgen,

Figure 5-9. *Changes in Value of $1,000 Investment,*
Ten Initial Public Offerings (IPOs), 1986–96

Value of investment (dollars)

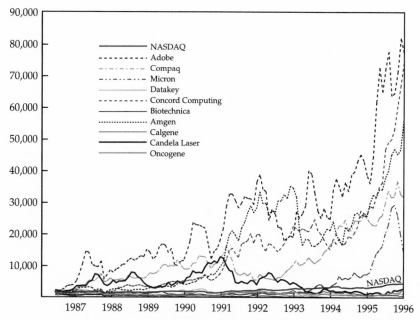

Source: Based on F. M. Scherer, Dietmar Harhoff, and Joerg Kukies, "Uncertainty and the Size Distribution of Rewards from Technological Innovation," *Journal of Evolutionary Economics* (forthcoming, 1999).

or Compaq Computer, on the other hand, one's initial $1,000 investment would have increased by the end of 1995 by thirty-two to seventy-eight times. At that time, the most successful five companies accounted for half the total equity value of all surviving companies plus the reinvested proceeds of twenty-three merged companies. Quite generally, as company performance and its valuation in the marketplace evolve over the course of fifteen years, the distribution of ultimate values becomes more and more skewed. In the aggregate, investment in the 131 companies yielded an end-of-1995 portfolio value nearly four times the initial assumed investment of $131,000 and approximately the same amount as investors would have realized had they invested their funds in a portfolio exactly tracking the NASDAQ index.[36]

For decades, the high-technology venture capital provision industry and the countless enterprises it nurtured were the United States' "secret weapon," imparting to U.S. industry a technological dynamism unmatched by other nations. But gradually the secret leaked out. Many nations began striving to build their own high-technology venture financing networks and to develop equity markets accommodating small and new enterprises.[37] The United Kingdom led the parade with $3.4 billion of venture capital disbursements in 1995, followed by France with $1.1 billion, Germany with $871 million, the Netherlands with $611 million, and Italy with $331 million.[38] Early investment patterns in these countries were less high-technology-oriented than in the United States. More than half of total European venture investments in 1995 went for buyouts and replacement capital. Seed investments amounted to only 0.6 percent of the total; startup investments, 5.2 percent; and expansion investments, 41 percent.[39]

In Canada venture capital firms were reported to be managing a capital pool (not an annual flow, as in the previous estimates) of some C$5.9 billion in 1995.[40] The share of total venture capital invested in technology-based firms is reported to have increased from 25 percent in 1989 to nearly 60 percent in 1994. However, the venture funds are said to invest mainly in larger companies and in tranches seldom smaller than C$750,000, leaving a financing gap for smaller startup enterprises.

In sum, important developments are under way in major industrialized nations to encourage entrepreneurial innovation by small technology-based companies and to supply them the financial wherewithal needed for their emergence and growth. The sinews that bound Prometheus are being unleashed, with possibly profound consequences for economic growth.[41]

Table 5-3. *Government Expenditures in Support of R&D,*
Selected Countries, 1993

Country	Total government outlays on R&D (billions of U.S. dollars)	Total intramural outlays (billions of U.S. dollars)	Percent of all R&D financed by government
Canada	3.41	1.58	40.1
United Kingdom	8.48	3.03	32.7
United States	69.88	17.10	37.7
France	13.61	5.58	43.5
Germany	14.92	5.54	36.7
Japan	12.30	6.94	21.6

Source: Organization for Economic Cooperation and Development, *Main Science and Technology Indicators*, no. 2 (Paris: 1996), tables 62, 54, and 35. See also U.S. National Science Board, *Science & Engineering Indicators: 1998*, p. A-179.

Public Policies toward Financing Innovation

The difficulties private enterprises experience in appropriating the benefits from their innovation investments and the risks associated with those innovations leave a place for government to intervene, either by enhancing private firms' incentives or supporting innovation directly through public funds. Both approaches are pursued. Debate centers on how, and how much, governments should intervene.

Direct Spending and Subsidies

An overview of the direct spending role of governments emerges from Organization for Economic Corporation and Development (OECD) statistics on various categories of government expenditures to support R&D, defined in a standardized way (see table 5-3). This analysis focuses on six leading industrialized nations reporting total government outlays on R&D, the amount spent in-house by government entities, and the fraction of all national R&D expenditures financed by government. The United States towers above the other nations in the absolute level of government outlays, more by virtue of its very large defense and space programs, mostly conducted in industry, than by proportionately larger involvement in financing the totality of national R&D investment.[42] Among the six, the French government finances the largest fraction of total national expenditure, the Japanese government the smallest fraction.

Despite substantial increases over recent decades in government support for basic research, the sufficiency of that support continues to be debated. This is a question that cannot be resolved here. Basic research, like some other laudable claims on public resources, has characteristics of a bottomless pit. Within the scientific communities, even fiercer disputes occur between proponents of spending on "big science" projects (such as the Superconducting Supercollider in the United States, similarly large-scale atom-smashing projects at CERN near Geneva, manned space stations, nuclear fusion, and sequencing the human genome) and those who would prefer to see the large blocks of resources megaprojects devour spread over a much wider array of investigations. Clearly, "big science" projects have merit. But individual nations (most notably, the United States) have tended to seek the lion's share of glory for their home-grown scientists rather than cooperating with other nations to spread the costs of gigantic projects. This was the case, for example, with the U.S. Supercollider, most of whose goals, it was argued, could be attained less expensively by allocating additional resources to the international facilities at CERN. It was only when the Supercollider project began exhibiting serious managerial problems, manifested in part by large prospective budget overruns, that congressional support for the project was withdrawn. Much smaller amounts were then appropriated to support cooperative research at CERN. An important U.K. white paper on science policy argues that "the largest scientific programmes require the investment of resources at a level which demands justification in world rather than national or European terms" and reports that in recent years, the share of the United Kingdom's science budget devoted to large science was deliberately reduced.[43]

Many nations have provided targeted subsidies to encourage applied industrial research and development, especially in areas having the potential to enhance national champions' competitive advantage in international trade both generally and in key future technologies. A danger exists that such R&D subsidy wars can escalate to a point at which serious distortions of international trade follow. To curb them, the Uruguay Round treaty signed at Marrakech in 1994 specified explicit limits for R&D subsidization; if these were overstepped, countervailing duty actions might be invoked. Government subsidies up to 75 percent of "industrial research" outlays are allowed, with "industrial research" defined as "planned research or critical investigation aimed at discovery of new knowledge, with the objective that such knowledge may be useful in developing new products, processes or services, or in bringing about a significant improvement to existing products, processes or services."[44] For "precompetitive" development outlays including conceptual design work and extending through the first prototype stage, subsidies up to 50 percent are permitted. Excluded from subsidy are R&D outlays for periodic alteration of existing products and processes and other continuing operations.

Tax Policy

Tax policy is another important instrument through which governments influence private industry's R&D investment decisions. There are four principal incentive-affecting vehicles as well as a host of special features too complex to enumerate here.[45]

First, even though spending for research and development is an investment in the future, most industrialized nations permit corporations to write off their outlays as current expenses, just as they allow the expensing of advertising and oil wildcatting. Canada, unlike the United States, also allows companies to write off capital expenditures for R&D (for instance, for laboratory construction and equipment). Even for new companies without current profits against which to offset R&D write-offs, current expensing is usually preferable unless loss carry-forward opportunities are meager.[46] A modest amount of industrial R&D in the United States has been organized under R&D partnerships, which allow wealthy sponsoring individuals to write off the outlays against income tax rates that were higher at the margin than the 35 percent rate applied (since 1993) to corporate income, or the zero rate applied to corporations without current profits or the expectation of profits sufficiently soon to make use of all loss carry-forwards.

Second, differential taxation of capital gains as compared with other classes of individual income can have significant incentive effects, especially for investments in new high-technology ventures. In the United States, changes in the tax laws giving preference to capital gains—with maximum individual capital gains tax rates falling in steps to 28 percent in 1977 and 20 percent in 1981—are said to have triggered substantial increases in venture capital investment.[47] It is difficult, however, to sort out the influence of those tax law changes from generally rising rates of return in venture funds at the time and changes in "prudent investor" rules governing the investment behavior of pension funds. The removal of most capital gains preferences between 1986 and 1997 probably had an incremental discouraging effect. Yet a record-setting boom in high-technology venture investments followed nevertheless during the 1990s, propelled by, among other things, massive inflows from pension funds whose capital gains are, at least until they are paid out to beneficiaries, exempt from taxation.[48] The extraneous influences contaminating these U.S. tax "experiments" are sufficiently complex that, even though most economists believe taxing current income and capital gains equally discourages venture investment, it is difficult to know whether the positive effects are sufficiently large to trump demands for budget-balancing and tax equity.

Third, investment in conventional plant and equipment is encouraged in some nations by allowing accelerated depreciation.[49] Increased physical plant investment in turn raises the sales and profits of capital goods suppliers, inducing them to perform more R&D.[50] Although the investment–R&D link is fairly well understood, there has been no definitive research on the more remote links between tax stimulants to investment on the one hand and induced capital goods R&D on the other.

Finally, many nations (but not the United Kingdom) provide special, explicit tax credit incentives for research and development investment. In this respect Canada has been a pioneer.[51] As early as 1962, it allowed corporations to deduct from their taxable income not only their actual outlays on R&D and R&D equipment, but also 50 percent of the difference between their current R&D expenditures and their 1961 expenditures. This provision was ended in 1966 but restored from 1978 to 1984. In addition, Canada initiated in 1977 a system of research and development tax credits, permitting (with certain complications) companies to take a direct credit against taxes for eligible *increases* in their R&D outlays. The magnitude of the credit varied from 10 to 25 percent of the R&D increments, with small firms and (until 1994) laboratories located in the Atlantic provinces receiving higher percentage credits than large companies performing R&D in other parts of Canada. In 1984, Canada again eliminated the extra R&D tax deduction but increased the tax credit ratios to between 20 and 35 percent (the higher number for small companies). The credit now applies to all eligible R&D outlays, not merely to incremental outlays, but the credit itself is subject to taxation at the corporate income tax rate. From a survey of Canadian companies, Edwin Mansfield estimated that the various R&D tax credits and augmented deductions increased corporate R&D spending by approximately 2.6 percent in 1982, or roughly

C$50 million.[52] The loss of revenue to the Canadian federal treasury was estimated at C$130 million.

Beginning in 1981, the United States emulated Canada by enacting an R&D tax credit law. As in Canada, there have been time gaps in the law's reenactment and frequent changes in the detailed provisions, diluting the credit's incentive effects.[53] At first a 25 percent tax credit was allowed on increases in qualifying company-financed R&D outlays over those of the preceding year. Later, the R&D base against which increases were measured was taken as the average of the previous three years, and still later (in 1989), as the average of the fixed base years 1984–88. The purpose of granting credits only on *increases* in spending is to maximize companies' incentive for a given tax sacrifice by government. But when each spending increment raises the base against which subsequent years' R&D outlays will be compared, a negative feedback effect is generated—hence the change in 1989 to a fixed historical base. In 1987 the credit rate was reduced to 20 percent, which continues to apply. The most careful econometric study of the impact of U.S. tax credit laws has been carried out by Bronwyn Hall.[54] Analyzing the R&D performance of some 1,000 corporations between 1980 and 1991, she found that the credit provisions raised company-financed R&D by approximately $2 billion a year (at 1982 price levels) while imposing a tax revenue sacrifice of roughly $1 billion. Total company-financed R&D in the United States during this period (in current year's price levels) rose from $30.5 billion in 1980 (1.1 percent of gross domestic product) to $90.6 billion (1.6 percent of GDP) in 1991. Uncertainty concerning the law's year-to-year reenactment undoubtedly reduced its incentive effects during the 1990s.

Patent Policy

Neglecting several other public policy instruments, I return to intellectual property. Patents and (for software) copyrights, we saw earlier, have traditionally been granted to retard the inroads of imitators and to help innovators recover their investments in research and development. How broad and strong such property rights should be continues to evoke controversy. For most industrialized nations, changes in domestic patent laws have in recent decades been mainly of a fine-tuning nature. Small changes, however, can sometimes have an important impact.

In the United States, the creation in 1983 of a centralized appellate court for the federal circuit, with responsibility for all patent-related appeals, unified decisionmaking, as Congress intended. But in addition, with a panel of judges drawn preponderantly from the patent law bar, the Court rendered decisions greatly strengthening the presumption of patent validity and broad scope in contested cases and increased to occasional billion-dollar thresholds the amount of damages awarded when infringement is proved. These changes bolster incentives for innovators in one respect. But they also make innovation more dangerous—indeed, much like walking through a mine field—in technologies with complex and overlapping patents of uncertain scope. The net effect on incentives is neither obvious nor known.

In Canada, patent law amendments passed during the late 1980s in anticipation of the Canadian–U.S. free trade treaty curbed what had been a widespread governmental practice of granting low-royalty compulsory patent licenses to companies offering generic substitutes for new drugs whose patents were still in force. The consequences include a strengthening of drug R&D incentives for multinational corporations, mostly located outside Canada; an agreement by those multinationals that they would move substantial amounts of R&D into Canada; and implementation by the government of drug price–control mechanisms as a substitute for compulsory licensing in the hope of containing health care costs.[55] A Canadian government study attempting to assess the balanced effect of these partly conflicting policies was under way in 1997.

The most important patent policy development of recent years came in provisions of the Treaty of Marrakech requiring less developed nations (LDCs), which for the most part had relatively weak patent laws, to upgrade their protection to the standards of industrialized nations within five years (or ten years for pharmaceutical product patents). For multinational corporations, the result will almost surely be enhanced profits on the sale of technologically advanced products in the third world and more rational siting of production facilities (since nations will no longer be able to deny patent protection on inventions not produced locally). The 64-billion-dollar question is whether those changes stimulate enterprises in LDCs to become more innovative in their own right, because patent protection on local inventions will be strengthened and because local firms will have a heightened incentive to invent around foreign firms' patents. Preliminary evidence suggests a negative prognosis,[56] but the longer-run consequences are hard to assess.

Human Capital 6

B Y FAR the most important input into the process of advancing science and technology is the creative power of human beings or, in economic terms, human capital. As we saw in chapter 3 from the investigations of Derek de Solla Price, the amount of human effort devoted to science and technology appears to have been rising at 4 to 5 percent a year over the past two or three centuries. Enormous advances in knowledge, technology, and material standards of well-being have resulted. Whether scientific and technological effort must continue growing at similar rates to maintain a steady rate of productivity growth—in other words, whether we are on a treadmill and must run ever faster to get ahead—is an open question that can hardly be resolved here. However, we can at least examine how much human capital is devoted to the science and technology enterprise and what growth trends it has exhibited in recent years. We can inquire also into the sources of likely future growth.

Current Effort and Trends

Figure 6-1 shows for twenty prominent nations the intensity with which scientific and engineering talent has been used in research and development activities (including R&D in both the private and public sectors), measured for the most recent year of data availability, relative to total national population.[1] Japan's role as the most powerful technological challenger to nations that achieved industrialization earlier is dramatically evident. Norway's status as the third most intensive user of R&D talent, trailing only Japan and the United States, is surprising. The United Kingdom and Canada occupy intermediate positions relative to the industrialized nation leaders. For nations in the bottom half of the distribution, the link between low levels of economic development and talent allocations seems clear.

Figure 6-2 provides a more dynamic view, tracing how the employment of scientists and engineers in research and development, measured relative to the size of national work forces, has grown in seven leading industrialized nations. Throughout most of the period following World War II, the United States was the most intensive allocator of human capital to R&D functions. Starting from far behind after the war's end, Japan moved into a virtual tie with the United States in 1989 and then jumped ahead. Whether Japan will retain its leadership despite recent economic setbacks remains to be seen. The United Kingdom stands out among the seven for its complete lack of growth until 1992. For Germany, the impact of reunification between East and West in 1990 is visible.

The United States exhibits retarded growth in the late 1980s and then a decline in the early 1990s. A similar decline occurred in total company-financed research and development outlays (dotted series in figure 6-3), adjusted to a constant-dollar basis using the gross domestic product deflator. The long-term growth trend, averaging 4.9 percent a year, is traced by the solid line. (A straight line on a

Figure 6-1. *Scientists and Engineers Engaged in R&D*
per 10,000 Population, Selected Countries, Various Years

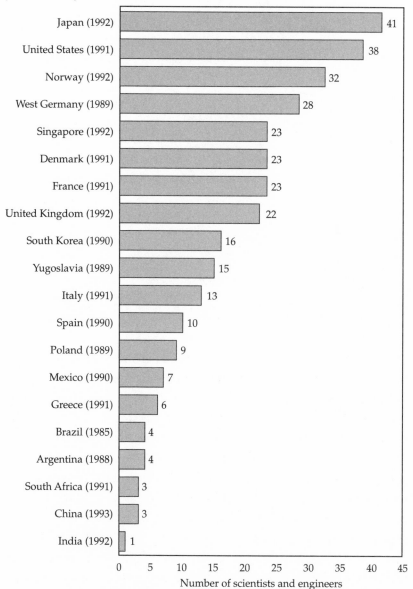

Number of scientists and engineers

Source: U.S. National Science Board, *Science & Engineering Indicators: 1996* (Washington: 1996), pp. 3–25.

Figure 6-2. *R&D Scientists and Engineers per 10,000 in Work Force, 1978–94*

Number per 10,000 in work force

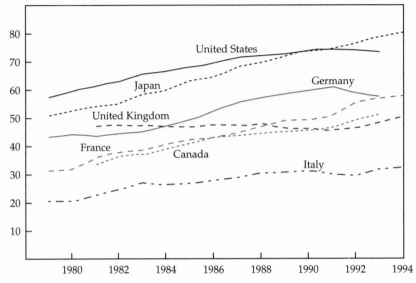

Source: U.S. National Science Board, *Science & Engineering Indicators: 1998* (Washington: 1998), appendix table 3-15, p. A-117.

logarithmic scale depicts a constant rate of growth.) Over the longer period for which comparable data are available, several cycles are evident. The average growth rate was 6.30 percent from 1953 to 1970, 3.75 percent during the 1970s (a decline accompanied by retarded productivity growth), 5.27 percent during the 1980s, and 3.64 percent during the first six years of the 1990s. Stagnant expenditures in the first four years of the 1990s, probably attributable to a recession and declining defense budgets, were followed by a brisk recovery to near-trend values. A more detailed industry-by-industry analysis for the years 1990–94 reveals that inflation-adjusted R&D outlays fell in eleven of the twenty-six disaggregated industry groups for which data were available.[2] Constant-dollar outlays fell by 9.8 percent between 1990 and 1994 in companies with 25,000 or more employees and by 5.8 percent in com-

Figure 6-3. *Trends in U.S. Company-Financed R&D Performance, 1953–96*

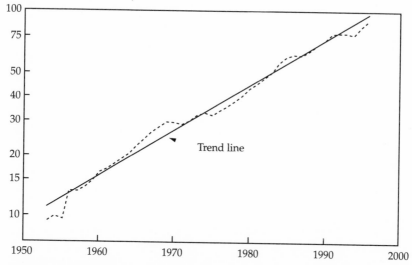

Billions of 1987 inflation-adjusted dollars

Source: U.S. National Science Foundation, *Research and Development in Industry: 1994* (Washington: NSF Report 97-331, 1997), p. 17; and U.S. National Science Board, *Science & Engineering Indicators: 1998*, p. A-121.

panies with from 10,000 to 24,999 employees—size categories in which National Science Foundation surveys have maintained consistent and continuous coverage. Firms in the smaller size categories reported substantial spending increases, but the reliability of the data is uncertain because the Foundation expanded its sample coverage in those categories during the comparison period.

Entrants into the Scientific and Engineering Work Force

Growth or retardation in the share of the work force devoted to research and development activities could occur because of changes in the demand for R&D talent or because of changes in the

Figure 6-4. *Trends in First University S&E Degrees for Six Nations, 1975–92*[a]

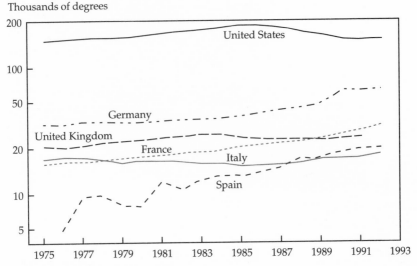

Thousands of degrees

Source: U.S. National Science Foundation, Special Report NSF 96-316, *Human Resources for Science and Technology: The European Region* (Washington: 1996), pp. 89–90.

a. U.K. data for 1992 are excluded because of an inconsistency, with graduates of polytechnic institutes being included for the first time that year.

supply of talent. Acknowledging that supply adjusts with a lag to changes in demand, I focus initially on the supply side.

According to data assembled by the U.S. National Science Foundation, higher education in science and engineering grew rapidly after 1975 in Europe, in part due to the creation and expansion of institutions such as polytechnics in the United Kingdom, Fachhochschulen in Germany, and instituts universitaires de technologie in France. The number of natural science degrees conferred in western European nations increased from roughly 54,000 in 1975 to 108,000 in 1991; the number of engineering degrees, from 51,000 to 106,000.[3] The education of scientists and engineers also expanded rapidly in east Asia—more so in the Asian tiger nations than in Japan and India, which sustained their most rapid growth during the 1950s. South Korea's graduate count increased by 4.5 times from 1975 to 1992; Taiwan's increased by 89 percent.[4] Canada experienced little increase during the 1980s. In

Figure 6-5. *Trends in U.S. S&E Bachelor's Degrees, 1960–94*

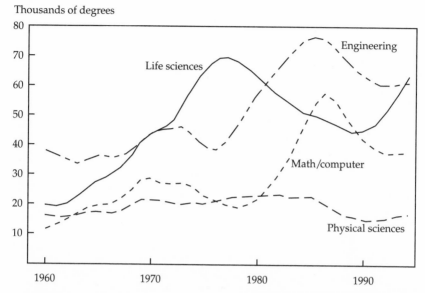

Source: U.S. National Science Foundation, *Science and Engineering Degrees: 1966–94*, NSF Report 96-321 (Arlington, Va.: 1996), especially tables 5 and 56; supplemented for 1960–65 by National Science Foundation, *Science and Engineering Personnel: A National Overview*, NSF Report 310 (Washington: 1990), table B-19; *Science & Engineering Indicators: 1998*, p. A-64; and (for population demographics) the *Statistical Abstract of the United States*.

the United States, the number of science and engineering graduates actually fell from a peak level in 1986, at least in part because of smaller numbers in the relevant age cohorts due to waning of the postwar baby boom. These and other trends in university baccalaureate science and engineering degree receipts are graphed for five European nations and the United States in figure 6-4. Spain is seen to have experienced especially rapid growth, surpassing Italy's degree awards after 1987.

A finer-grained view reveals much about the changing popularity over time of scientific and engineering studies in the United States. Figure 6-5 presents raw counts of the number of baccalaureate degrees awarded in various R&D–related fields from 1960 through 1995. All exhibit declines in the 1980s, although from widely varying peak years. Degree receipts peaked earliest in the

Figure 6-6. *U.S. S&E Degrees per 1,000 Twenty-Two-Year-Olds, 1960–94*

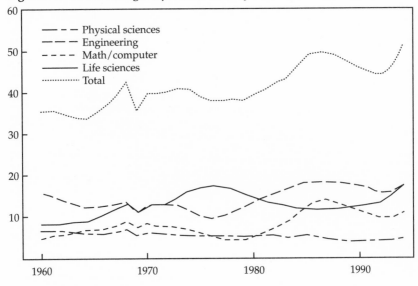

Source: See figure 6-5.

life sciences (which include agricultural sciences along with substantial numbers of premedical students), but there was a sharp
rebound in the early 1990s, perhaps attributable at least in part to
the revolution in industrial biotechnology. Computer sciences and
mathematics had the most recent peak, but again, despite the
tremendous spread of computers into American offices and households, experienced a sharp decline beginning in 1986. Studies in the
physical sciences (physics, chemistry, and earth sciences) grew
slowly and then declined more abruptly.

These movements over time were influenced among other things
by demographic trends, notably the upsurge in university enrollments as baby boomers born during the late 1940s and early 1950s
completed their secondary studies and as an increasing fraction of
eligible individuals sought university training. Figure 6-6 corrects
for the raw population effect, relating first science and engineering
degree awards to the number of twenty-two-year-olds. It becomes
clear that virtually all of the pronounced growth during the late

1960s and 1970s was attributable to demographic changes. Despite primary and secondary education reforms instigated in the wake of the 1957 Sputnik shock and claims that the Apollo moon landing program would fire young people's imaginations and lead them to choose careers in science and technology, there was no notable upward trend during the 1960s and 1970s in the number of individuals per 1,000 twenty-two-year-olds who elected to pursue science and engineering (S&E) education.[5] The most pronounced growth occurred in the life sciences, with the weakest links to the space program. Increased S&E matriculation during the early 1980s was propelled in part by a massive expansion of military research and development (increasing the demand for engineers) and by the microcomputer revolution (increasing interest in, or demand for, computer science graduates). Both booms faded in the late 1980s.[6] The 1990s upturn for all S&E degree awards combined was driven largely by the increase in life science degrees.

Figure 6-7 provides still another perspective, relating the number of bachelor's degrees in science and engineering fields to the total number of university degrees awarded at the baccalaureate and first professional levels.[7] Figure 6-6 showed a flat trend; figure 6-7 reveals a declining trend during the 1960s and early 1970s as the number of business, law, and medical degrees soared—between 1960 and 1990, for instance, from 4,643 MBA or equivalent degrees to 76,676; from 9,240 law degrees to 36,485; and from 7,032 medical degrees to 15,075. To be sure, many MBA recipients and some newly anointed lawyers obtained their undergraduate training in science or engineering, but embarked thereafter on a path that would take them out of the laboratory and into the office. This evidence supports at least for the United States an OECD group's conclusion that "Australia, Canada, New Zealand, Spain and the United States stand out for their relatively low proportion of graduates in engineering, which is seemingly correlated with these same countries having above-average proportions of graduates in law and business studies."[8]

Figure 6-7. *U.S. S&E Degrees as Percentage of Bachelor's and First Professional Degrees, 1960–94*

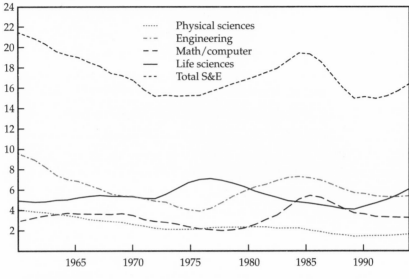

Source: See figure 6-5.

In contrast to these mostly pessimistic trends, there is evidence of improvement in the intellectual ability of students who elect undergraduate science or engineering studies. The fraction of U.S. college freshmen with A– or better high school grade averages choosing undergraduate S&E programs increased from 35 percent in 1985 to 43 percent in 1995.[9] Since students beginning S&E programs composed 18 percent of all freshmen in 1985 and 17 percent in 1995, it seems clear that S&E studies attracted a disproportionate share of the *best* university entrants.

Figure 6-8 traces on a logarithmic scale the number of doctoral degrees awarded in the natural sciences (including mathematics and computer sciences) and engineering in four nations between 1975 and 1992. All four exhibit growth over time—Japan from the lowest start, at 1,245 degrees in 1975 to 2,318 in 1992.[10] U.K. doctorates grew from 3,592 to 5,177; German doctorates from 3,575 to 8,804. As in the United Kingdom and Germany, the growth of

Figure 6-8. *Trends in S&E Doctorates, Four Nations, 1975–92*

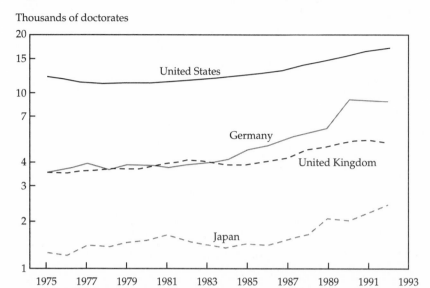

Source: U.S. National Science Board, *Science & Engineering Indicators: 1996,* appendix tables 2-28 and 2-30. The later data in *Science & Engineering Indicators: 1998,* pp. A-73 and A-83, are not fully comparable.

degrees in the United States accelerated during the late 1980s, rising to 17,802 in 1992 and 19,219 in 1995. However, much of that growth is attributable to an influx of foreign students. The number of S&E doctorates received by U.S. citizens and permanent residents rose only from 8,995 in 1977 to 10,553 in 1992 and 13,076 in 1995. An omen of possible trend reversal is suggested by the decrease in total U.S. S&E graduate student enrollments from a peak of 292,668 in 1993 to 279,998 in 1995.[11]

Why Not S&E Careers?

There are apparently powerful influences limiting the expansion of scientific and engineering work forces once the participation

Table 6-1. *Median Annual Salaries, April 1995, for U.S. Work Force*
Participants Receiving Degrees during 1993 and 1994

	U.S. dollars	
Field of study	Bachelor's degree	Master's degree
All engineering	33,500	44,000
Computer and mathematical sciences	30,000	43,200
Life and related sciences	22,000	31,200
Physical and related sciences	25,000	35,000
Social and related sciences	21,000	30,000

Source: U.S. National Science Foundation, Science Resources Studies Division, *Data Brief*, "Recent Engineering Graduates Out-Earn Their Science Counterparts," November 8, 1996.

levels found in the most highly developed nations are achieved, and sometimes at even lower levels. Why might this be? And why were U.S. efforts to interest more university entrants in science or engineering careers unsuccessful?

Career choices are molded by an individual's intrinsic ability and interest, educational experiences at the primary and secondary levels, prospects that one will secure satisfying employment, and expected financial rewards in the chosen vocation. From a short-run perspective, the financial rewards that follow receipt of science or engineering degrees cannot be said to be unattractive. The median annual salaries reported in April 1995 by U.S. work-force participants who received various kinds of degrees during 1993 and 1994 are shown in table 6-1. Engineers fared best among recipients of both bachelor's and master's degrees; computer scientists were next best. Life science graduates fared worst among the S&E degree holders, which may explain in part why their numbers experienced such a precipitous decline beginning in 1985; see figures 6-6 and 6-7. However, they were better off than students who had majored in social sciences, who in turn probably received better-paying jobs on average than students in the humanities, for whom data were unavailable. Recipients of master's degrees in business administration are apparently excluded from the "social and related sciences" category. In 1994, the *average* starting salary (with

Table 6-2. *Average Annual Incomes of U.S. Males,*
Selected Professional Specialties, 1992

Field	U.S. dollars
Manufacturing administrators and officials	60,643
Finance, insurance, and real estate administrators	66,589
Accountants and auditors	40,172
Engineers	46,638
Natural scientists and mathematicians	45,202
Postsecondary teachers	39,872
Primary and secondary teachers	29,820
Lawyers and judges	89,756
Health diagnosticians (physicians)	93,454

Source: U.S. Bureau of the Census, *Money Income of Households, Families, and Persons in the United States: 1992,* Current Population Reports Series P60-184 (1993), pp. 152–53. Later editions of the same report provide less occupational detail.

bonuses) for graduates of the top twenty U.S. MBA programs, as rated by *Business Week*, was $87,024—nearly twice the starting salary of a median engineering master's degree recipient.[12] MBA recipients from the top twenty university with the *lowest* average starting salary in 1994 began at an average of $54,720—well above the average for *all* engineering master's recipients. The reason for the explosion of MBA enrollments at the expense of S&E careers seems clear.

In the long run (although not as long as Keynes's long run), further changes materialize. Nations organize their statistics on income distribution in heterogeneous ways, so the relevant data for the United States and the United Kingdom are addressed separately. The average 1992 annual incomes of U.S. males engaged in a representative sampling of professional specialties are reported in table 6-2. Table 6-3 shows the average *weekly* earnings in 1996 for U.K. male professionals whose pay was unaffected by absences. In both nations, engineers and scientists (and the university teachers they might alternatively become) receive much lower salaries on average than high-level managers, attorneys, and physicians. To

Table 6-3. *Average Weekly Earnings for Male Professionals Whose Pay Was Unaffected by Absences, United Kingdom, 1996*

Field	British pounds
General managers, large companies and organizations	1,733
Treasurers and company financial managers	936
Bank, building society, and post office managers	662
Computer systems and data processing managers	653
Engineers and technologists	504
Biological scientists and biochemists	521
Other natural scientists	485
University and polytechnic teaching professionals	577
Legal professionals	727
Medical practitioners	853

Source: Office for National Statistics, *New Earnings Survey: 1996*, Part D: Analyses by Occupation (London: 1996).

be sure, many engineers and (especially in the pharmaceutical and biotechnology industries) scientists leave the laboratories and join top management, enjoying as a result substantially higher salaries. In making that move, they largely forsake the nuts-and-bolts functions of advancing technology, although they may still play crucial roles in directing the innovative process. How important taking an early career jump and obtaining an MBA or equivalent degree is to making that transition is not known; for the United States, it is probably quite common except in the most science-oriented enterprises. And the averages suggest that for a bright, ambitious young person, a more promising general path to affluence is to study law or medicine.

For those who are willing to take vows of poverty and chastity (but never obedience!), life as an academic scientist or teacher of engineering exerts a considerable attraction. And unless outstanding individuals are drawn into teaching, it is questionable whether superbly trained graduates will emerge from the educational pipeline.[13] Here, too, problems exist. A perspective on the U.S. situation, with which the author is most closely acquainted, illustrates. The professors who in an earlier era would have retired from teaching in 1997 were born in 1932—a year ushering in a six-year period with the lowest birth rates the United States experienced during the twentieth century. Thus relatively few university professors will reach traditional retirement ages in the next decade. U.S. universities expanded their faculties rapidly during the 1960s and 1970s to accommodate the surge of postwar baby boom students. New faculty members hired at age twenty-five in 1970 will reach the traditional retirement age in the year 2010. But recent changes in U.S. law have made it impossible for universities to force professorial retirement on account of age, which means that many faculty members will choose to continue teaching until they reach age seventy or later. All this means that there will be relatively few vacancies for faculty replacement during the next fifteen years. This situation will be ameliorated only slightly by renewed enrollment growth and modest expansion requirements as the children of postwar baby boomers enter colleges and universities, gradually raising the number of U.S. residents in the eighteen to twenty-four age cohort from 24.6 million in 1996 to 30.1 million in 2010.[14] Given severe pressures on college and university budgets, much of this growth is likely to be absorbed through increases in "productivity"—that is, larger class sizes—rather than augmented faculty appointments. Certainly, the opportunities for rewarding careers in teaching are not going to be as abundant as they were during the 1960s and 1970s. One consequence is proliferation and prolongation of postdoctoral fellowships, if they are available, or low-paying jobs as transient teachers when they are not.[15] During the mid-1990s, National Institutes of Health funds were sufficient to support only the top 11 percent of applicants for postdoctoral fellowships in microbiology and genetics.

Teachers of science and technology face special problems. At first-rate universities and colleges, faculty members are expected also to carry out research. Indeed, their effectiveness as mentors for tomorrow's young scientists and engineers demands continued research vitality. But in many areas of science, the sharply rising cost of running a serious laboratory has intensified the competition for funds from (mostly) federal research support agencies. To operate a laboratory in molecular biology, for example, expenditures for apparatus, reagents, vectors, and research assistance will run to at least $200,000 a year and more commonly $300,000 a year. Many budding young academicians are reluctant to enter a life of incessant fund-grubbing, and, given plausible limitations on grant support, few will be able to succeed at the game even when they are willing to play. As a result of this long-run outlook, the shortage of tenure-track teaching positions, and the short-run prospect of protracted research assistantships or postdoctoral fellowships at low pay, an unknown but probably substantial number of the United States' best and brightest will choose careers other than academic science.

Some will elect completely different disciplines, such as medicine, law, or business. Others will stay the course and eventually enter jobs in industry. The industry alternative is especially viable in microbiology, and the world will be enriched by the wonder drugs and other products that result. But in other fields, the prospects are much bleaker. Largely because of fluctuations in defense and space spending, the industrial demand for engineers and most physical scientists in the United States has always been cyclical. Demand boomed during the missile and space race of the late 1950s and early 1960s; it slumped when Vietnam War budgetary pressures forced a reallocation of funds from high-technology weapons to personnel, guns, and ammunition; it rose again in the renewed cold war buildup initiated by President Carter and sustained by President Reagan; and during the early 1990s, as the perceived cold war threat faded, it declined briskly. The combination of atrophied academic openings, falling demand for military R&D, and modest growth of civilian-sector R&D will together severely constrict job openings in technology and science, leading many young university students to pursue careers in fields other than science and engineering.

For those who have completed their university training and are unable to find new or replacement jobs in science or technology, difficult readjustments will have to be made.[16] But most such individuals have generous endowments of human capital, and experience with previous science and technology job recessions suggests that on the whole they will adapt well.[17] A U.S. National Science Foundation survey found that only 52.8 percent of the undergraduate engineering degree holders, 10.0 percent of the life sciences degree recipients, and 22.1 percent of the physical science bachelor's recipients were working in the same field in 1993 as their undergraduate training. Those who worked in unrelated non–S&E fields numbered 19.4, 37.0, and 25.2 percent respectively.[18] For the recipients of doctoral degrees in those three specialties, the proportions working in the same field were 59.9, 57.4, and 57.4 percent respectively; those working in unrelated non–S&E fields numbered 12.6, 19.8, and 9.2 percent.

A robust economy helps scientists and engineers unable to secure employment directly in their specialties to obtain challenging alternative jobs. For university-trained individuals who worked for the formidable defense and space complex of the former Soviet Union, on the other hand, budgetary cutbacks and the unavailability of funds to pay persons still employed have been catastrophic.[19] The inability of a poorly functioning economic system to reallocate resources efficiently has propagated a massive waste of human capital.

From Where Will Future Growth Come?

At present, the principal human capital problem on a world-wide scale appears to be absorbing in productive scientific and engineering work the highly trained individuals already on hand. In the long run, however, the challenge is likely to be different. If Price's findings about past S&E growth can be extrapolated into the future, the number of scientists and engineers will have to increase substantially. It is unknown whether exponential increases in effort are needed merely to keep the frontiers of science and technology advancing at a roughly constant pace. But it is clear that if the benefits of modern technology are to be diffused to members of the world's population who do not yet enjoy them in any measure because they live in underdeveloped nations, there will have to be a massive expansion of S&E talent. From where is that growth to come?

Table 6-4 uses UN survey data to provide an overview of where in the world scientists (including those in the agricultural sciences, but excluding medical sciences) and engineers were being educated during 1992. Because distinctions between programs leading to university degrees and those that stopped short of a degree (granting a certificate) were incomplete, the focus is on all programs, whether they led to a degree or only a certificate. From partial statistics, I estimate that focusing on degree programs only would reduce the numbers by approximately 21 percent. The first three columns are drawn from a sample of sixty-five nations, mostly relatively large, that accounted for 80 percent of the world's population and for which the data were reasonably complete. Extrapolations are then made in the last two columns for the entire world on the basis of less complete data.[20]

Perhaps the most striking fact is that more than half the world's population resides in poverty-ridden nations, that is, nations with gross national product (at estimated purchasing power parity) of less than $2,000 per capita, with a very low intensity of science and engineering education. Whereas the weighted average number of students in science and engineering programs per 100,000 national residents was 387 for all sixty-five sample nations, the comparable number was 105 for nations with average GNP per capita below

Table 6-4. *Overview of World Science and Engineering Training, 1992*

GNP per capita	Number of nations	S&E students per 100,000 population	S&E students (millions)	S&E students adjusted for undercount (millions)	Percent of world population
More than $12,000	21	801.6	6.40	6.45	14.5
$5,000–$11,999	21	764.5	6.47	7.45	18.3
$2,000–$4,999	12	395.6	1.69	3.71	16.3
Less than $2,000	11	105.0	2.44	2.74	50.9
All nations	65	386.7	17.00	20.35	100.0

Source: United Nations Economic and Social Council, *World Education Report: 1995* (Oxford: 1995), tables 1, 8, and 9.

Table 6-5. *Nations Housing Highest Number of S&E Students, 1992*

Country	Millions of S&E students
Russia	2.40
United States	2.38
India	1.18
China	1.07
Ukraine	0.85
South Korea	0.74
Germany (united)	0.73
Japan	0.64
Italy	0.45
Philippines	0.44

Source: United Nations Economic and Social Council, *World Education Report: 1995* (Oxford: 1995), tables 1, 8, and 9.

$2,000.[21] Clearly those nations are severely handicapped relative to their wealthier brethren in achieving technology-based economic development. Quite generally, the more prosperous a nation's citizens, the larger the number of science and engineering students per 100,000 population.[22]

Altogether, roughly 20 million of the world's university-level students were specializing in science and engineering during 1992. Of these, more than two-thirds were in the wealthiest nations— those with GNP per capita of $5,000 or more.

A somewhat different light is shed on the subject by arraying nations according to the number of science and engineering students they housed. Table 6-5 shows the top ten nations in 1992, with the number of students expressed in millions.[23] Two points leap out. First, although India and China educate a relatively small fraction of their population, those countries have such large populations that they are among the world's top four S&E educators. The future of science and technology is going to depend importantly on how well their blossoming human capital is used. Second, at least in 1992,

Table 6-6. *Nations with Highest Numbers of Science and Engineering Students per 100,000 Population*

Country	S&E students per 100,000 population
South Korea	1,701
Ukraine	1,639
Russia	1,619
Finland	1,421
Canada	1,104
Taiwan	1,013
Ireland	957
United States	933
Germany (united)	904
Chile	879
Israel	866
Spain	860
Unweighted average for 65 nations	555

Source: United Nations Economic and Social Council, *World Education Report: 1995* (Oxford: 1995), tables 1, 8, and 9.

Russia and Ukraine were turning out huge numbers of technically trained individuals for jobs that did not exist. Unless the main fragments of the former Soviet Union bring order to their economic houses, there will be even more massive human capital waste.

Table 6-6 gives still another perspective, listing the twelve nations (from the main sample of sixty-five) with the highest numbers of science and engineering students per 100,000 population.[24] Again, we see in Russia and Ukraine the potential for enormous waste. A reason for the rapid and (at least, until late 1997) successful transition to high-technology industrialization by South Korea and Taiwan becomes apparent. Strong science and engineering education programs also help explain (along with generous corporation income tax preferences) the recent movement of high-technology industry toward the Republic of Ireland.

Tapping the Untapped Talent Pool

Roughly half of the world's population is located in the least developed nations, where a winnowing process has already identified more than 2 million students promising enough to be encouraged to study science or engineering. Sampling from such a large population allows an appreciable chance, assuming adequate standards of maternal and child nutrition, that a von Neumann, a Curie, a James Watson, a James Watt, or some combination of their rare genius could emerge. Also, in the nations of the former Soviet Union there is a pool of student talent from which, at least in the past, some of the world's great mathematicians (Lobachevski and Pontryagin), physicists (Kapitsa), and chemists (Mendeleyev) have emerged. How can that latent human capital be cultivated?

Most of it will be needed at home if China and India move onto the technological development paths that Japan, Korea, Taiwan, and other Asian tiger nations have pursued in recent decades. It will also be needed at home if Russia and its spin-offs evolve a conducive institutional framework and adopt economic policies that permit their long-suffering citizens to benefit from capitalism's economic potential, foreseen so vividly by Karl Marx. The "ifs" here are monumental; future history will be significantly influenced by them. But whatever happens, there are also opportunities and perhaps obligations for the leading industrialized nations to ensure that outstanding talent does not go to waste.

A few Western governmental and private organizations have taken advantage of the crisis in science in the former Soviet Union by issuing contracts for work in Russia on specific scientific or technological tasks.[25] Contracts have been let on a much wider scale for software development in India, with interface requirements flowing over the Internet to the Indian software writers and the completed programs flowing back at the speed of light.[26]

Table 6-7. *Degrees in S&E Disciplines Awarded to Foreign Students, 1992*

| | United States | | United Kingdom | |
Field	Number	Percent	Number	Percent
First (baccalaureate) degrees				
Natural sciences	4,556	4.3	1,013	5.3
Engineering	4,582	7.3	1,914	19.5
Doctoral degrees[a]				
Natural sciences	4,863	32.8	1,020	26.7
Engineering	3,249	50.5	719	49.7

Source: U. S. National Science Board, *Science & Engineering Indicators: 1996* (Washington, 1996), p. 67.
a. The U.S. doctoral degree data are for 1993 rather than 1992.

Immigration is a more traditional alternative. In the past, the United Kingdom, the United States, and Canada have done extraordinarily well in welcoming able young scientists and engineers from other lands and encouraging them to give vent to their creative energies. James Watson was an American post-doc in England when he worked with Francis Crick to identify the structure of DNA and its implications for heredity; John von Neumann was a refugee from Hungary when he conceived the basic principles of digital computer architecture; the era of atomic weapons and atomic power was ushered in by refugee scientists such as Rudolph Peierls and Otto Frisch in England and Leo Szilard, Enrico Fermi, George Kistiakowski, Eugene Wigner, and Edward Teller in the United States. The opportunity still exists to draw immigrants who will make major contributions.

U.S. and British universities attract many students from foreign countries into their S&E programs. Degrees awarded to foreign students and their percentage relative to all students' degrees in S&E disciplines in 1992 are shown in table 6-7. For both nations, doctoral programs in engineering exert an especially strong pull on foreign students.[27] Foreign students' presence in U.S. advanced science and engineering degree programs continued to rise during the 1990s.

In 1995, fully 40 percent of the 26,515 U.S. S&E doctorate recipients were citizens of other nations, compared with 27 percent in 1985.[28] Of the 10,493 nonnative doctoral degree recipients, 7,660 came from Asia, 1,253 from Europe, 505 from parts of North America other than the United States (including 274 Canadians), 424 from Africa, and 358 from South America. Among the Asian recipients were 2,751 from the People's Republic of China, 1,239 from Taiwan, 1,204 from India, and 1,004 from South Korea. Large-scale enrollment of Chinese students in U.S. universities began in early 1979 after a surprise decision by Deng Xiaoping in July 1978 to allow and indeed to encourage study abroad.

What happens to foreign students after they have completed their degrees? Much depends on employment conditions in their native lands, the students' desire to integrate themselves into new and different cultures, and host nation immigration policies. Canada has maintained an especially open immigration policy. U.S. policy has fluctuated over the years, but the Immigration Act of 1990 permitted large increases in employment-based quotas for highly skilled individuals.[29] As a result, the admission of scientists and engineers with permanent visas jumped from 12,659 in 1990 to 22,871 in 1992.[30] However, a sharp decline followed between 1993 and 1995.[31] The influence of home market conditions and culture is shown by sizable differences in the percentage of 1992 U.S. doctoral degree recipients who planned to remain in the United States (see table 6-8). The percentages for those with *firm* plans to remain in the United States are shown in parentheses.

Japan and Korea, with robust opportunities at home in engineering, had particularly modest rates of planned immigration. France, with a life-style difficult to replicate in the United States, had the next lowest rate in engineering and the lowest rate in the natural sciences. Holding general preferences constant, students from nations in which English is widely spoken had more success in landing the jobs that firmed up their plans to remain in the United States.

Table 6-8. *Percentage of Foreign-Born 1992 U.S. Doctoral Degree Recipients Planning to Remain in the United States*[a]

National origin	Natural science degrees	Engineering degrees
China	89.6 (54.2)	87.1 (36.2)
Taiwan	58.1 (35.2)	50.2 (22.5)
Japan	52.0 (34.0)	24.0 (12.0)
South Korea	46.5 (30.8)	29.9 (14.9)
India	84.3 (60.2)	82.9 (55.2)
United Kingdom	68.6 (52.9)	82.4 (64.7)
Canada	55.8 (44.2)	58.7 (43.5)
Germany	54.5 (39.0)	55.6 (38.9)
France	41.0 (28.2)	40.0 (16.0)
Spain	75.0 (44.4)	50.0 (33.3)

Source: U. S. National Science Board, *Science & Engineering Indicators: 1996* (Washington, 1996), appendix table 2-34.

a. Percent having firm plans to remain in the United States is shown in parentheses.

A generally hospitable U.S. stance toward talented immigrants has meant that not only new graduates, but also previous graduates, add to the U.S. science and engineering human capital stock. A survey of doctoral recipients, old and young, in 1993 revealed that immigrants composed substantial fractions of all U.S. residents with S&E doctoral degrees.[32] Some took their doctorates in the United States; others migrated to the United States only after receiving degrees elsewhere. The fractions of all relevant doctorate holders in each of these categories and both together are shown in table 6-9. Altogether, 23.0 percent of all science and engineering degree holders (including the social sciences) under the age of seventy-five residing in the United States were foreign born.[33] Immigrants made up an even larger fraction—27.8 percent—of the doctorate holders engaged in research and development as their primary or secondary work activity.

Thus, in the past and at present, offering student slots to individuals born in other nations and integrating those persons into the creative work force has been a significant contributor to the stock of

Table 6-9. *Percentage of U.S. Resident Doctoral Degree Holders in 1993 Who Were Immigrants*

Field	U.S. doctorate	Foreign doctorate	Combined
Engineering	33.0	7.2	40.2
Computer/math sciences	25.9	7.7	33.6
Physical sciences	15.9	10.0	25.9
Life sciences	10.6	10.7	21.3

Source: U. S. National Science Board, *Science & Engineering Indicators: 1996* (Washington, 1996), appendix table 3-14.

human capital in nations such as the United States, the United Kingdom, and (less well documented, but true) Canada. How aggressive the recruitment of foreign-born students and R&D professionals should be in the future is an important policy issue. Clearly, migrating individuals exercising free choice over the venues in which they work and study believe they gain through relocation. The nations receiving the direct benefit of their efforts are also strengthened technologically. But the nations from which talented scientists and engineers emigrate lose an important resource. If they provide an environment in which the would-be émigrés could do productive work, and especially if the émigrés' contributions to the alleviation of underdevelopment at home are sacrificed, one must be apprehensive about encouraging the so-called "brain drain."[34] But if the alternatives to emigration are third-best vocational options and thwarted realization of human potential, a strong case can be made for invigorated immigration policies.

Other Supply-Side Determinants

Three additional issues—the underuse of women in science and engineering, weaknesses in primary school education, and duplication of scientific efforts—must be discussed more briefly.

Table 6-10. *Percentage of Nonacademic Science and Engineering Jobs Held by Women, 1980–90*

Field	1980	1990
Physical scientists	16.2	22.7
Life scientists	25.6	32.5
Computer/math scientists	25.9	36.2
Engineers	4.4	9.2
Social scientists	38.1	51.4

Source: U. S. National Science Board, *Science and Engineering Indicators: 1996* (Washington, 1996), p. 3-13.

Women are an important potential resource thus far tapped insufficiently in the industrialized nations. As prejudices erode and large numbers of women enter the professional work force, they can be expected to augment the ranks of professional scientists and engineers. The changes that occurred over the course of one decade are shown in table 6-10.[35] UN statistics reveal generally similar proportions of women enrolled in natural science, engineering, and agricultural science degree programs (but excluding medical sciences) at the university level across diverse nations in 1992. Some representative female S&E student percentages for individual nations are shown in table 6-11. The average for fifty-three nations with available data was 24.2 percent, although some cultural biases are apparent. African nations had the lowest average at 20.0 percent, Asian nations the second lowest at 22.3 percent, and South and Central American nations the highest average at 28.4 percent.

Statistics for the United States were not reported, but National Science Foundation surveys reveal that in 1995, women obtained 45.5 percent of the baccalaureate degrees in the natural sciences (including mathematics and computer sciences) and 17.3 percent of the engineering degrees, or 41.7 percent combined—figures that signal future increases in the female S&E work force ratio.[36]

One factor limiting young students' interest and ability to perform well in science and engineering is the weakness of science and

Table 6-11. *Female Candidates in Natural Science, Engineering, and Agricultural Science Degree Programs, Selected Nations, 1992*

Country	Percent
Canada	20
United Kingdom	24
Germany	25
France	34
Italy	33
Switzerland	14
Japan	10
Korea	14
India	27
Iran	15
Saudi Arabia	25
Nigeria	9
Brazil	31

Source: United Nations Economic and Social Council, *World Education Report: 1995* (Oxford: 1995), table 9.

mathematics education at pre-university school levels, and especially in elementary schools. Here increasing opportunities for women in professional jobs constitute a two-edged sword. A half century ago, when discrimination against women in science and business was rife, many highly intelligent, well-educated women who chose full-time employment became primary and secondary school teachers.[37] As opportunities outside teaching opened up, and as salaries in other professions rose relative to teachers' salaries, a pronounced decline occurred, at least in the United States, in the academic ability and attainments of new teachers. Richard Murnane and his colleagues report that in 1967 college graduates with measured IQ scores of 130 were nearly as likely to enter teaching as were graduates with (population-average) scores of 100.[38] By 1980, the more numerous cohort of graduates with scores of 100 were on average four times as likely to enter teaching as those with scores of 130. In 1993, only 7 percent of grades 1–4 science teachers and 32 percent of grades 5–8 teachers had completed university majors or minors in science or science education.[39] For mathematics teachers, 7 percent of grades 1–4 teachers and 18 percent of grades 5–8 teachers had comparable backgrounds in the study of mathematics. If students enter secondary school, where the ratios of teachers with specialized university training are much higher, with inadequate facility in, or aversion to, science and especially mathematics, it is difficult to turn them around and direct them toward careers in science or engineering.

As the world's scientific and engineering work force grows, despite the obstacles we have identified, diminishing returns could set in if the investigators in any given nation merely duplicate the R&D of groups located in other nations, rather than thrusting out in new and original directions. This is a possibility, but there is reason to believe that it may not be a serious problem. In less developed nations, much of the augmented S&E work force should be occupied implementing technology already brought to an advanced level in the most progressive nations. Full diffusion of existing technology can hardly occur unless engineers are working to implement it at virtually every significant production site.

In more creative endeavors, the likelihood of duplication is higher. This is not necessarily bad. When uncertainty is substantial and the best approach to solving a scientific or technical problem cannot be identified, Mao's admonition, "Let one hundred flowers bloom," is wise strategy. The more important, either economically or intellectually, is the problem addressed, the larger is the optimal number of independent research approaches.[40] It would be much less desirable if individual nations consciously emulated projects, and especially large-scale projects, for essentially chauvinistic reasons or to free themselves from the need to import high-technology products developed elsewhere. A prominent example was the costly uncoordinated development in recent decades of quite similar military fighter aircraft in the European NATO nations, Sweden, Israel, and the United States (not to mention the Soviet Union, where more plausible grounds for independent development existed). Especially for big-ticket high-technology products, there is much to be said for letting international trade flourish. The competition (since terminated) between the United States and Europe in constructing ultra–high energy particle accelerators is another example. In smaller-scale scientific projects, the best assurance against excessive and wasteful duplication is to maintain scholarly traditions of openness and early publication of significant research results. The rapid growth of electronic mail networks will facilitate such communication, which will help the invisible hand guide investigators to a tolerably efficient allocation of research resources.

Conclusion | 7

ECONOMIC GROWTH has bestowed a remarkable bounty on citizens of the world's industrialized nations. If anything, the growth of real income per capita has been more rapid during the past half century for most developed and developing nations than it was in earlier periods. Can progress be sustained at anything like the rates experienced since World War II by nations near the technological frontier? Or will a retardation of growth be imposed by diminishing returns—if not on land and in using the earth's mineral products, then on investment and the technological innovations that enhance its productivity, or on the earth's ability to absorb the environmental byproducts of progress?

This book has emphasized technological change as the most crucial dynamic force leading to economic growth. A key question for the future, therefore, is whether a brisk pace of technological advance can be sustained.[1]

There is a centuries-old tradition of gazing with wonder at recent technological achievements, surveying the difficulties that seem to thwart further improvements, and concluding that the most important inventions have been made and that it will be much more difficult in the future to achieve comparable rates of advance. Such views have always proved to be wrong in the past, and there is no reason to believe that they will be any more valid in the foreseeable future. Scientists and engineers engaged in hand-to-hand struggle with the unsolved problems of microbiology, medicine, communication, transportation, nutrition, housing, improvement of humans' mental capabilities, the use of increased leisure, and much else are convinced that important strides remain to be achieved. More worrisome are the historical statistics assembled by Derek de Solla Price and his projection into the future.[2] For nearly two centuries the amount of effort devoted to science and, more recently, to formally organized research and development activities appears to have been growing at an average annual rate in excess of four percent. Population increases much more slowly once nations have passed the demographic transition associated with economic development. Must scientific and engineering effort continue rising at more than four percent annually to keep per capita incomes growing in technologically advanced nations at historically realized rates—between 1894 and 1994, at 1.38 percent in the United Kingdom, 1.92 percent in the United States, and 2.08 percent in Canada?[3] If so, we are headed for trouble, since, as Price observes, in less than a century "we should have two scientists for every man, woman, child, and dog in the population."[4]

The detailed analysis of the U.S. experience in chapter 6 reveals that there are significant limits to expanding the scientific and engineering work force in a frontier-straddling nation. It is unclear whether the key constraints lie on the supply or the demand side. On the supply side, improvements in education, especially at the primary and secondary levels, might augment the number of people choosing careers in science and engineering. A more direct but partly conflicting increase could come from erosion of long-standing barriers to such careers for women—a change that hinges in no small measure on innovations in the way science and mathematics are taught in the schools. Even more important would be narrowing the substantial differentials between the salaries paid scientific and engineering employees and those received by physicians, attorneys, and managers. But the genesis of such a change would have to come from the demand side. Are attorneys and managers paid much more on average than engineers and scientists because the former are more productive in some fundamental way, or do the differentials emerge from deep-seated cultural traits leading managers to reward kindred souls more liberally and impelling business enterprises to proliferate zero-sum legal contests in which the party with the better attorney wins?[5] Altering the status quo will not be easy, although strenuous efforts to do so must be commended. A demand-side opportunity with greater current momentum is the reduction, as the cold war winds down, of allocations of scientific and engineering talent into the development and production of weapon systems. The extent to which those resources are reallocated into activities contributing more directly to long-term economic growth will depend in part on salary responses in private-sector research and development occupations. If one consequence of reduced military R&D employment is further depression of relative scientific and engineering pay scales, university graduates are likely to opt increasingly toward pursuing nontechnical careers, and the technological progress dividend from ending the cold war will disappoint.

Many of these same tendencies operate, even if at attenuated levels, in other highly industrialized nations of the world. One thing that distinguishes the United States is its vibrant high-technology venture capital institutions, financing the emergence of countless new enterprises offering a diversity of new technical approaches and the willingness to engage in a Darwinian struggle determining which ones best meet the test of the marketplace. The rewards are skewed and the risks are high. But unlike lotteries that attract widespread participation in industrialized nations, despite actuarial payoff expectations only half the value of funds risked,[6] venture capital is for investors a positive-sum game, yielding average payoffs well in excess of the amounts invested. Technological progress will almost surely be accelerated as other nations successfully emulate U.S. venture capital approaches. To succeed, they need to cultivate three groups: a class of investors who understand the rules, risks, and rewards of the game; financial intermediaries who are at ease in the world of technology as well as in the domains of money and management; and a class of technologists willing to accept the risk of advancing new ideas in which they have faith. Crucial to the emergence of the third group is the spread of recognition within the business community and the larger public that failures are inevitable, that failing is no disgrace, and that those who fail should be given a chance to start again. For some nations, the required changes border on revolutionary. In Germany, for example, where I have lived for many years, the requisite attitudes existed after World War II because virtually everyone had directly experienced failure. But as the economic miracle matured and prosperity was shared by new generations unaccustomed to failure and old generations fearful of relapse, tolerance for failure atrophied.

Advances in basic scientific knowledge enrich the opportunities for making specific market-oriented technological advances. Historically, the lags from basic discoveries to commercial products or processes have ranged from ten to forty years, but the intervals have almost surely been declining as the links between the scientific and technological communities have tightened.[7] To stimulate commercial innovation activities but neglect basic science is like installing a high-speed pump while letting the water in a reservoir dwindle. Because it is difficult to appropriate the discoveries that flow as a public good from scientific research, private profit-oriented (as distinguished from philanthropic) motives for supporting basic science are deficient. If governments do not fill the gap, it is unlikely to be closed. The governments of North American and European nations have for the most part been generous in providing such support. In the United States, for example, federal government backing for basic science is estimated to have risen from $1.15 billion (in constant 1987 dollars) in 1953 (three years after the formation of the National Science Foundation) to $13.31 billion in 1995—in other words, at an average annual growth rate of 5.8 percent.[8] Industrial basic research expenditures rose at an average annual rate of 5.0 percent from $696 million in 1953 (again, at 1987 price levels) to $5.67 billion in 1995. It is difficult politically to sustain such growth rates for an activity performed by an elite minority of the nation's citizens—although support is stronger for medically oriented research, whose benefits legislators understand intuitively, than for research in other areas. Ideally, a long-run non-partisan consensus should be achieved on maintaining a stable rate of real growth in government funding—for example, at 4 to 5 percent a year. Even then, difficult conflicts would have to be resolved periodically on allocations to specific disciplines and between "big science" and "little science." Resolving such questions requires responsible priority-setting by elected leaders of the principal scientific associations.

The political problem grows in complexity as more nations, such as Japan and Korea, expand their financing of basic research. Just as business enterprises are inclined to free-ride on the research spillovers from other enterprises and nonprofit institutions, so nations may free-ride on other nations' support of science. Until World War II, the United States was more of a free rider than an independent contributor. After the war, Japan joined the ranks of leading free riders, but its government now recognizes that greatly increased domestic scientific efforts are necessary for sustained technological vitality. The free-rider problem might be alleviated in part by an agreement among the leading industrialized nations to maintain minimum rates of real basic research growth, with the minimums being greater, the smaller the fraction that a nation's basic research outlays are to gross domestic product. For "big science" projects, increased cooperative cost-sharing among the leading nations is highly desirable. National governments should also recognize that generously subsidizing study abroad and overseas travel for collaboration and meeting attendance enhances the productivity of the world's research effort by facilitating knowledge flows, ensuring through direct competition that high standards prevail, and reducing unwarranted duplication of basic research projects.

The data analyzed in chapter 6, and especially table 6-4, reveal that the most important opportunity for sustaining growth of the world's scientific and engineering effort lies in tapping more effectively the vast pool of talent being educated in Asia and the former Soviet Union. That talent can be used indigenously (and indeed must be if economic development is to proceed) or in a more broadly integrated world framework. In either case, help from the leading industrialized nations is needed.

Internally, the less developed nations encounter three main barriers to effective use of scientific and technical talent for their rapid economic development. Most important is the lack of, or critical shortcomings in, a legal and institutional framework that encourages vigorously independent risk-taking and dynamic competition or, in Schumpeterian terms, "creative destruction" among business enterprises. Second, and closely related, is a scarcity of business entrepreneurs willing and (by virtue of education and experience) able to take advantage of the opportunities for development offered by modern technology. Third, because the less developed nations have by definition low real per-capita incomes, the exigencies of keeping body and soul together for their citizens make it difficult to devote substantial funds to research and development activities whose benefits accrue only after appreciable lags.

The core fragments of the former Soviet Union, with 16 percent of the world's science and engineering students (counting only

Russia and Ukraine), provide a poignant case in point. To put the matter bluntly, the vital task of transitioning from authoritarian socialism to free market capitalism has been badly botched.[9] The most basic mistake was attempting to implant capitalism into an environment lacking both appropriate legal and social institutions and a sizable cohort of individuals with genuine entrepreneurial skills (as distinguished from the ability to deal in *Blat*). An important component of the problem could have been alleviated had the Soviet Union actively encouraged during the 1970s the emergence of an independent small business sector adept at filling consumer needs left unsatisfied by clumsy central plans. That opportunity was missed, and so post-perestroika history began with deficient preconditions. As a result, the greatest rewards went to those who used political connections to transfer (rather than creating) wealth and to those who satisfied pent-up demands by importing consumer products rather than innovating and producing them at home. Inability of the governments to offer investors secure property rights and to collect taxes complicated matters. In the absence of radical reform (such as erecting barriers to consumer goods imports and external cash transfers on personal account, fragmenting still-cumbersome enterprises, redistributing ownership rights, and then guaranteeing subsequent property rights), it is likely to take decades, if it occurs at all, before the problems are corrected to provide a basis for dynamic economic growth.

The former Soviet bloc nations of eastern Europe, which had a shorter tradition of socialism and (in most cases) inherited stronger small-business sectors, have made better progress in large part because the preconditions were more favorable. China, too, seems well on the way toward developing a dynamic entrepreneurial business sector, despite the continuing overhang of inefficient state-owned, provincially protected enterprises.[10] And India, which at current population growth rates could become the world's largest nation, may yet succeed in reforming the governmental and cultural institutions that have held its business sector back.

The prosperous industrialized nations can help emerging and redeveloping nations move toward the technological frontier in a variety of ways. The most pressing need is implementation of existing modern technology in industry. That in turn requires investment capital (much of which can be raised domestically), active technology transfer mechanisms, and entrepreneurship capable of helping technically trained graduates do their important work. Foreign direct investment is the quickest way to achieve all three, but as recent repayments and exchange rate crises in southeast Asia testify, developing nations will be reluctant to rely predominantly on foreign capital sources. Indigenous enterprises must also be encouraged to move to the technological frontier. Assuming that the appropriate legal and regulatory reforms were in place, this could happen best if the wealthy industrialized nations provided a new and quite different version of Marshall Plan assistance—subsidizing with funds and talent the education of technology-oriented managers, supporting similarly the creation and growth of technology transfer institutions (such as the highly successful International Rice Research Institute, which contributed crucially to the Green Revolution in agriculture), and encouraging Western business enterprises to license their technology on favorable terms.

The need for rapid technology transfer will be particularly acute if the world's nations seriously pursue the environmental protection initiatives adopted at the Rio de Janeiro and Kyoto international conferences. Given traditional developmental trajectories, rapidly developing economies are likely to add the largest increments of pollutants to the earth's environment. But they also offer the richest untapped opportunities for adopting state-of-the-art pollution-control and energy-saving technologies right from the start, rather than on a more expensive retrofit basis. Helping them do so should be an early priority of internationally funded research, development, and technology transfer institutes.

At first transfer and (where necessary) adaptation of already existing technology would be the focus of the initiatives suggested here. But as business enterprises master existing technology, they and the technology institutes with which they cooperate would begin to perform research and development of a more genuinely innovative character, just as Japan began decades ago to pioneer new methods of shipbuilding and automobile manufacture and to devise superior new products such as high-quality point-and-shoot cameras, facsimile machines, fiber optical cable terminal equipment, and video cassette recorders.[11] The world's technological frontiers will be extended. And gradually the developing nations will begin also to make independent contributions to scientific knowledge.

In the meantime, some individuals from developing nations will be capable of performing path-breaking scientific work, given the proper environment. Ensuring that such talent realizes its potential depends in large measure on the most developed nations. Immigration policy is the obvious instrument. As chapter 6 observed, the leading industrial nations have done well in admitting foreign students into their universities, encouraging them, and absorbing them into their faculties and industrial laboratories. Preferential access of academically talented individuals should continue.[12] The number of students able to carry out graduate studies in Western universities depends significantly on the willingness and ability of their home governments to finance them.[13] More could be done, and the pace of scientific advance would be accelerated marginally, if Western governments increased the financial support available for meritorious foreign students.

Few of the initiatives suggested here will be without cost or popular among politicians' domestic constituencies. But there is much at stake—no less than a continuation of the scientific and technological progress that has done so much during the past two centuries to advance human welfare.

Notes

Chapter 1

1. See especially David Landes, *The Unbound Prometheus: Technology Change and Industrial Development in Western Europe from 1750 to the Present* (Cambridge University Press, 1969); and Joel Mokyr, *The Lever of Riches* (Oxford University Press, 1990).

2. On the methods used to measure real income at the national level and the difficulties encountered, see Angus Maddison, *Monitoring the World Economy: 1820–1992* (Paris: OECD, 1995), pp. 117–47. On the sizable underestimates that can occur for the satisfaction of a basic need such as illumination, see William D. Nordhaus, "Do Real Output and Real Wage Measures Capture Reality?" in Timothy F. Bresnahan and Robert J. Gordon, eds., *The Economics of New Goods* (University of Chicago Press, 1997), pp. 29–70.

3. Continuous compounding is assumed. For Korea, the time spans are 1912–93.

4. See, for example, Edward F. Denison, *Accounting for Slower Economic Growth: The United States in the 1970s* (Brookings, 1979); Martin N. Baily and Alok Chakrabarti, *Innovation and the Productivity Crisis* (Brookings,

1988); Nestor E. Terleckyj, *Changing Sources of U.S. Economic Growth: 1950–2010* (Washington: National Planning Association, 1990), pp. 42–43; F. M. Scherer, "Lagging Productivity Growth: Measurement, Technology, and Shock Effects," *Empirica*, vol. 20 (1993), pp. 5–24; and Charles I. Jones, "The Upcoming Slowdown in U.S. Economic Growth," Working Paper 6284 (National Bureau of Economic Research, November 1997).

5. See, for example, the symposium on "New Growth Theory" in the *Journal of Economic Perspectives*, vol. 8 (Winter 1994), pp. 3–72; and Jan Fagerberg, "Technology and International Differences in Growth Rates," *Journal of Economic Literature*, vol. 32 (September 1994), pp. 1147–75.

Chapter 2

1. See, for example, John Maynard Keynes, *The General Theory of Employment, Interest and Money* (London: Macmillan 1936), chapter 23 (on mercantilism and theories of under-consumption). Keynes credits August Heckscher with the "two birds" observation.

2. Adam Smith, *An Inquiry into the Nature and Causes of the Wealth of Nations* (1776; New York: Modern Library edition, 1937), p. 11.

3. Ibid., p. 9.

4. Ibid., p. 10.

5. Ibid., p. 321.

6. Clifford F. Pratten, "The Manufacture of Pins," *Journal of Economic Literature*, vol. 18 (March 1980), pp. 93–96.

7. David Ricardo, *The Principles of Political Economy and Taxation* (1817); Thomas Robert Malthus, *Principles of Political Economy* (1820).

8. The curve chosen for illustration has the equation $TQ = 180L - 0.6\,L^3$, where total output TQ and total labor L are measured in millions of units (bushels per year and person-years). The marginal product curve is $dTQ/dL = 180 - 1.8\,L^2$.

9. In mathematical terms, it is the integral of the marginal product function $180 - 1.8\,L^2$.

10. The higher wage will cause a restriction of employment on the lands of the original British Isles to point K, at which the new higher wage of 125 coincides with the M_o curve.

11. Ricardo recognized that the subsistence wage was in some respects socially determined and that workers might adjust to generous wages by

choosing higher standards of living rather than increased family sizes. But, Ricardo argued in a revision to his treatise, "although this might be the consequence of high wages, yet so great are the delights of domestic society, that in practice, it is invariably found that an increase in population follows the amended condition of the labourer." *The Principles of Political Economy and Taxation* (London: Dent & Sons, 1911), p. 278.

12. Compare John Eatwell and others, eds., *The New Palgrave Dictionary of Economics* (Macmillan 1987, vol. 1, p. 371 (article by Murray Milgate).

13. U.S. Bureau of the Census, *Historical Statistics of the United States: Colonial Times to 1957* (Washington: 1960), p. 72; and *Economic Report of the President*, February 1996, p. 316.

14. Yujiro Hayami and Vernon W. Ruttan, *Agricultural Development: An International Perspective*, rev. ed. (Johns Hopkins University Press: 1985), pp. 131, 467–71.

15. Why some nations are so far from the production frontier and how nations move toward the frontier will be explored in depth in chapter 4.

16. Where dY/dt is the growth rate of real income, e_Y is the income elasticity of demand, and $d(Q/L)/dt$ is the growth rate of labor productivity, the growth rate of the labor force share of a sector is roughly $((1 + dY/dt) e_Y) / (1 + d(Q/L)/dt)$.

17. The decline in the share of the manufacturing and mining labor force was also associated in part with a growing U.S. trade deficit in merchandise, but that effect amounts to less than 3 percentage points out of the nearly 20 percentage point decrease shown by figure 2-5.

18. Donella H. Meadows and others, *The Limits to Growth: A Report for the Club of Rome's Project on the Predicament of Mankind* (New York: Universe Books, 1972). Between 1970 and 1990, the growth rate slowed to 1.9 percent. Continued growth at 1.9 percent a year would imply world populations of 9.4 billion in 2020, 16.6 billion in 2050, and 29.3 billion in 2080. On a continuation of the debate at a meeting of the American Anthropological Society, see "Will Humans Overwhelm the Earth? The Debate Continues," *New York Times*, December 8, 1998, p. D5.

19. See, for example, Harold Barnett and Chandler Morse, *Scarcity and Growth: The Economics of Natural Resource Availability* (Johns Hopkins University Press, 1963).

20. See Kingsley Davis, "The World Demographic Transition," *Annals of the American Academy of Political and Social Science*, vol. 237 (January 1945), pp. 1–11.

21. Roy F. Harrod, "An Essay in Dynamic Theory," *Economic Journal*, vol. 49 (March 1939), pp. 14–33; and Evsey Domar, "Capital Expansion, Rate of Growth, and Employment," *Econometrica*, vol. 14 (April 1946), pp. 137–47.

22. Robert M. Solow, "A Contribution to the Theory of Economic Growth," *Quarterly Journal of Economics*, vol. 70 (February 1956), pp. 65–94.

23. Robert M. Solow, "Technical Change and the Aggregate Production Function," *Review of Economics and Statistics*, vol. 39 (August 1957), pp. 312–20.

24. Because other economists had achieved the same insight previously using less sophisticated mathematical methods. See, for example, Richard R. Nelson, "How New Is Growth Theory?" *Challenge*, September-October 1997, pp. 29–58.

25. Perhaps the most ambitious statistical decomposition of Solow's residual into components such as improved education, on-the-job experience, changes in the labor force's gender mix, the effect of the business cycle on the use of capital and labor, changes in the pattern of resource allocation (for example, with shifts from agriculture into industry), the effects of government regulation, and improvements in technology was undertaken by Edward F. Denison. See his *Accounting for Slower Productivity Growth* (Brookings, 1979).

Chapter 3

1. Karl Marx and Friedrich Engels, *The Communist Manifesto*, ed. Samuel H. Beer (New York: Appleton-Century-Crofts, 1955), p. 12.

2. Ibid., p. 14.

3. Joseph A. Schumpeter, *The Theory of Economic Development*, trans. Redvers Opie (New York: Oxford University Press, 1961). The first editions of *The Theory of Economic Development* mentioned Marx at best peripherally. Only in a new preface to the 1937 Japanese translation did Schumpeter acknowledge the parallelism of his views with those of Marx. It seems likely that if Schumpeter had admitted a Marxian influence in conservative *fin de siècle* Austria, his book would not have been received nearly as enthusiastically.

4. Ibid., pp. 88–89. For an illustration of the distinction drawn from a famous case, see F. M. Scherer, "Invention and Innovation in the Watt-

Boulton Steam Engine Venture," *Technology and Culture*, vol. 6 (Spring 1965), pp. 165–87.

5. Joseph A. Schumpeter, *Capitalism, Socialism, and Democracy* (New York: Harper, 1942).

6. Ibid., p. 66.

7. For a fiftieth-anniversary retrospective on this thesis, see F. M. Scherer, "Schumpeter and Plausible Capitalism," *Journal of Economic Literature*, vol. 30 (September 1992), pp. 1416–33.

8. Schumpeter encouraged and nurtured mathematical methods, even though he himself never mastered the necessary techniques sufficiently to be an innovator.

9. Robert M. Solow, "Perspectives on Growth Theory," *Journal of Economic Perspectives*, vol. 8 (Winter 1994), p. 52.

10. U.S. National Science Foundation, *Funds for Research and Development in Industry, 1958* (Washington: 1961).

11. The poem lines are not included in his paper, "Research and Development for the Emergent Nations," in Richard A. Tybout, ed., *Economics of Research and Development* (Ohio State University Press, 1965), pp. 422–37. The poem was included later in Richard P. Beilock, ed., *Beasts, Ballads, and Bouldingisms* (Transaction Books, 1980), p. 96. In the 1965 symposium article, at pp. 425–26, Boulding strikes a theme that will resonate in this book. Citing the discussion on "improvements in machinery" that is quoted in chapter 2 (note 4), Boulding observes, "Now . . . old Adam has put his finger on the heart of the matter. It is as we develop a special class of people engaged in research and development and, one would like to add, education, that we at last find a learning process whose horizon seems to be almost unlimited."

12. See especially his *Invention and Economic Growth* (Harvard University Press, 1966).

13. This conclusion received further support in a more comprehensive cross-sectional analysis of capital goods inventions covering 245 narrowly defined U.S. manufacturing industries. The higher the investment, the larger was the appropriately lagged number of linked capital goods inventions, with a correlation coefficient of +0.74. The relationship held both for inventions made by firms internally to meet their own process technology needs and for inventions made in specialized capital goods–supplying industries to serve other industries' needs. For raw materials inventions, the correlation with using industry purchases was much

weaker. F. M. Scherer, "Demand-Pull and Technological Innovation: Schmookler Revisited," *Journal of Industrial Economics*, vol. 30 (March 1982), pp. 225–38.

14. An important early contribution was Nestor Terleckyj, *Effects of R&D on the Productivity Growth of Industries: An Exploratory Study* (Washington: National Planning Association, 1974). For a review of the literature, see Zvi Griliches, "R&D and Productivity," in Paul Stoneman, ed., *Handbook of the Economics of Innovation and Technological Change* (Oxford: Basil Blackwell, 1995), pp. 52–89.

15. See also Jeffrey G. Williamson, "Productivity and American Leadership: A Review Article," *Journal of Economic Literature*, vol. 29 (March 1991), pp. 51–68.

16. See William J. Baumol, Sue Anne Batey Blackman, and Edward N. Wolff, *Productivity and American Leadership: The Long View* (MIT Press, 1989), p. 97; and Robert Barro and Xavier Sala-i-Martin, *Economic Growth* (McGraw-Hill, 1995), pp. 27, 420.

17. See, for example, his "Investment in Human Capital," presidential address to the American Economic Association, *American Economic Review*, vol. 51 (March 1961), pp. 1–17.

18. Robert E. Lucas Jr., "On the Mechanics of Economic Development," *Journal of Monetary Economics*, vol. 22 (July 1988), pp. 3–42. For an analysis of the instabilities that might ensue if returns were not precisely constant, see Solow, "Perspectives on Growth Theory," p. 50.

19. Paul M. Romer, "Increasing Returns and Long-Run Growth," *Journal of Political Economy*, vol. 94 (October 1986), pp. 1001–37; and (more important) "Endogenous Technological Change," *Journal of Political Economy*, vol. 98, supplement to no. 5 (1990), pp. S71–102.

20. John P. Foley, ed., *The Jefferson Cyclopedia*, vol. 1 (New York: Russell and Russell, 1967), p. 433.

21. A key equation in Romer's system is $dA/dt = \delta H_A A$, where dA/dt is the annual growth of the design knowledge stock, A is the level of the design knowledge stock, H_A is the amount of human capital allocated to creating new designs, and δ is a scaling parameter.

22. For a survey of the literature, see Zvi Griliches, "The Search for R&D Spillovers," *Scandinavian Journal of Economics*, vol. 94 (1992 Supplement), pp. 29–47.

23. See, for example, Kenneth Arrow, "Economic Welfare and the Allocation of Resources for Invention," in Richard R. Nelson, ed., *The*

Rate and Direction of Inventive Activity (Princeton University Press: 1962), pp. 619–22.

24. Matters become more complex if the new product cannibalizes the sales of other products on which positive price–cost margins were previously earned. See, for example, F. M. Scherer, "The Welfare Effects of Product Variety: An Application to the Ready-to-Eat Cereals Industry," *Journal of Industrial Economics*, vol. 28 (December 1979), pp. 113–34; and Steven S. Wildman, "A Note on Measuring Surplus Attributable to Differentiated Products," *Journal of Industrial Economics*, vol. 33 (September 1984), pp. 123–32.

25. Edwin Mansfield and others, "Social and Private Rates of Return from Industrial Innovations," *Quarterly Journal of Economics*, vol. 91 (May 1977), pp. 221–40.

26. The fraction of R&D originating in nonmanufacturing industries, and especially service industries, has since then increased—for instance, to approximately 25 percent in 1995. U.S. National Science Foundation, Science Resources Studies Division, *Data Brief*, December 16, 1997, p. 2. The central point of figure 3-3 remains—much of the benefit from R&D flows out to other using industries.

27. F. M. Scherer, "Inter-Industry Technology Flows and Productivity Growth," *Review of Economics and Statistics*, vol. 64 (November 1982), pp. 627–34. For application of a similar approach to a much larger set of nations, see Organization for Economic Cooperation and Development, *Technology and Industrial Performance* (Paris: OECD, 1997).

28. See Dale W. Jorgenson, "Productivity and Economic Growth," in Ernst R. Berndt and Jack Triplett, eds., *Fifty Years of Economic Measurement* (University of Chicago Press: 1990), pp. 19–118; Paul Krugman, "The Myth of Asia's Miracle," *Foreign Affairs*, June 1994, pp. 62–78; Alwyn Young, "The Tyranny of Numbers: Confronting the Statistical Realities of the East Asian Growth Experience," *Quarterly Journal of Economics*, vol. 110 (August 1995), pp. 641–80; and (for a more balanced view) Susan M. Collins and Barry P. Bosworth, "Economic Growth in East Asia: Accumulation versus Assimilation," *Brookings Papers on Economic Activity, 2:1996*, pp. 135–203.

29. This was recognized at an early date by Robert Solow in "Technical Progress, Capital Formation, and Economic Growth," *American Economic Review*, vol. 52 (May 1962), pp. 76–86.

30. See, for example (on the experience of the U.S. steel industry), F. M. Scherer, *Industry Structure, Strategy, and Public Policy* (HarperCollins, 1996),

pp. 182–92. When capital goods become obsolescent quickly, as in computers and semiconductor manufacturing, slow market growth is much less of an impediment to rapid adoption of new technology.

31. U.S. National Science Board, *Science & Engineering Indicators: 1998* (Washington: 1998), p. A-368.

32. Dietmar Harhoff, "Strategic Spillovers and Incentives for Research and Development," *Management Science*, vol. 42 (June 1996), pp. 907–25.

33. Leading examples include Adam Jaffe, "Real Effects of Academic Research," *American Economic Review*, vol. 79 (December 1989), pp. 957–70; James D. Adams, "Fundamental Stocks of Knowledge and Productivity Growth," *Journal of Political Economy*, vol. 98 (August 1990), pp. 673–702; Zoltan Acs, David Audretsch, and Maryann Feldman, "R&D Spillovers and Recipient Firm Size," *Review of Economics and Statistics*, vol. 76 (May 1994), pp. 336–40; Edwin Mansfield, "Academic Research Underlying Industrial Innovations," *Review of Economics and Statistics*, vol. 77 (February 1995), pp. 55–65; Lee Branstetter, "Are Knowledge Spillovers International or Intranational in Scope? Microeconomic Evidence from the U.S. and Japan," Working Paper 5800 (National Bureau of Economic Research, October 1996); Francis Narin and others, "The Increasing Linkage between U.S. Technology and Public Science," *Research Policy*, vol. 26 (December 1997), pp. 317–30; and Adam Jaffe and Manuel Trajtenberg, "International Knowledge Flows: Evidence from Patent Citations," Working Paper 6507 (National Bureau of Economic Research, April 1998).

34. The concept originated with Edward H. Chamberlin, *The Theory of Monopolistic Competition* (Harvard University Press, 1933), and experienced rich theoretical elaboration during the 1970s and 1980s.

35. Economics is surely the only discipline in which a scholar can win the Nobel Prize for proving the existence of that which plainly does not exist. I refer to the prizes awarded to economists for proofs of the "existence" (in the mathematical, not the observational, sense) of competitive general equilibrium.

36. This latter point is supported strongly in Wesley M. Cohen and Steven Klepper, "A Reprise of Size and R&D," *Economic Journal*, vol. 106 (June 1996), pp. 925–51.

37. But see Daniele Archibugi and Mario Pianta, *The Technological Specialization of Advanced Countries* (Dordrecht: Kluwer, 1992), who show that the smaller nations' aggregate R&D expenditures are, the more those efforts tend to be specialized over a narrow array of technologies.

38. See Gene M. Grossman and Elhanan Helpman, "Endogenous Innovation in the Theory of Growth," *Journal of Economic Perspectives*, vol. 8 (Winter 1994), p. 40; and Martin N. Baily and Hans Gersbach, "Efficiency in Manufacturing and the Need for Global Competition," *Brookings Papers on Economic Activity: Microeconomics, 1995*, pp. 307–47.

39. F. M. Scherer, *International High-Technology Competition* (Harvard University Press, 1992).

40. Grossman and Helpman, "Endogenous Innovation," p. 40.

41. See Howard Pack, "Endogenous Growth Theory: Intellectual Appeal and Empirical Shortcomings," *Journal of Economic Perspectives*, vol. 8 (Winter 1994), especially pp. 60–63.

42. For a survey of the relevant literature, see Richard E. Caves, *Multinational Enterprises and Economic Analysis*, 2d ed. (Cambridge University Press, 1996), pp. 166–88. For a diversity of views, see the symposium on Technology Transfer in the *Annals of the American Academy of Political and Social Science*, vol. 458 (November 1981). The most carefully controlled statistical investigation shows strong productivity effects for the plants of multinational companies but minimal or even negative effects for indigenous plants in the same industry. See Brian Aitken and Ann Harrison, "Do Domestic Firms Benefit from Foreign Direct Investment? Evidence from Panel Data," working paper, Columbia University, 1998.

43. See Larry Westphal, K. S. Kim, and Carl Dahlman, "Reflections on the Republic of Korea's Acquisition of Technological Capability," in Nathan Rosenberg and Claudio Frischtak, eds., *International Technology Transfer: Concepts, Measures, and Comparisons* (Praeger, 1983), pp. 167–221; and Alice Amsden, *Asia's Next Giant* (Oxford University Press, 1989).

44. Wesley M. Cohen and Daniel A. Levinthal, "Innovation and Learning: The Two Faces of R&D," *Economic Journal*, vol. 99 (September 1989), pp. 569–96.

45. See Barro and Sala-i-Martin, *Economic Growth*, pp. 434–40; Robert E. Hall and Charles I. Jones, "The Productivity of Nations," Working Paper 5812 (National Bureau of Economic Research, November 1996); the symposium on "What Have We Learned from Recent Empirical Growth Research?" *American Economic Review*, vol. 87 (May 1997), pp. 173–88; and World Bank, *World Development Report: 1997* (Oxford University Press, 1997) (whose theme is "The State in a Changing World").

46. Ishtiaq Mahmood, "Growth and the Puzzle of Political and Economic Equilibrium," working paper, Harvard University, 1996. See also

Robert J. Barro, "Determinants of Economic Growth: A Cross-Country Empirical Study," Working Paper 5698 (National Bureau of Economic Research, August 1996).

47. See, for example, his "Increasing Returns and Long-Run Growth," pp. 1019–20.

48. See also Julian Simon, *The Ultimate Resource* (Princeton University Press, 1981), pp. 198–210, who argues that with increases in the size of the world's population, and especially the population of technically trained individuals, there are more people to conceive new and original ideas, and, hence, once the embodiments of those ideas spread, a more rapid rate of technological progress.

49. See Scherer, *International High-Technology Competition*, pp. 178–81. However, a certain amount of "duplication" is desirable, the more so, the greater the technological uncertainty at the start of a project and the deeper the stream of benefits realizable through technical success. See F. M. Scherer, *Innovation and Growth: Schumpeterian Perspectives* (MIT Press, 1984), chapter 4.

50. Derek J. de Solla Price, *Little Science, Big Science* (Columbia University Press, 1963), p. 19.

51. Compare Scherer, *Innovation and Growth*, pp. 258–59.

52. For a similar argument emphasizing the improbability of continuing increases in both human capital and research and development efforts, see Charles I. Jones, "The Upcoming Slowdown in U.S. Economic Growth," Working Paper 6284 (National Bureau of Economic Research, November 1997).

53. Chapter 6 elaborates on facets of this issue.

Chapter 5

1. See Paula Stephan, "The Economics of Science," *Journal of Economic Literature*, vol. 34 (September 1996), pp. 1199–1262.

2. See Richard R. Nelson, "The Link between Science and Invention: The Case of the Transistor," in the National Bureau of Economic Research conference volume, *The Rate and Direction of Inventive Activity* (Princeton University Press, 1962), pp. 549–83; John Tilton, *International Diffusion of Technology: The Case of Semiconductors* (Brookings, 1971), pp. 74–77; and

Michael Riordan and Lillian Hoddeson, *Crystal Fire: The Birth of the Information Age* (Norton, 1997), pp. 163–200.

3. This discussion draws on Sheryl Winston-Smith, "The Cohen-Boyer Patents: A Case Study," John F. Kennedy School of Government term paper, Harvard University, January 1996.

4. See "Basic Research Is Losing Out As Companies Stress Results," *New York Times*, October 8, 1996, p. 1. For one exception, see "Microsoft Plans 300% Increase in Spending for Basic Research in 1997," *New York Times*, December 9, 1996, p. C1. The pharmaceutical and biotechnology industries have also actively increased their basic research efforts.

5. U.S. National Science Foundation, *Research and Development in Industry: 1994* (Washington: 1997), p. 54. Basic research is defined for National Science Foundation surveys as "original investigation for the advancement of scientific knowledge and which does not have specific immediate commercial objectives."

6. U.S. National Science Board, *Science & Engineering Indicators: 1998* (Washington: 1998), p. A-127.

7. U.S. Office of the President, Office of Science and Technology Policy, *The U.S. Technology Policy*, September 1990.

8. Chancellor of the Duchy of Lancaster, *Realising Our Potential: A Strategy for Science, Engineering and Technology* (London: HMSO, May 1993), p. 49.

9. See U.S. Department of Commerce, National Institute of Standards and Technology, *Advanced Technology Program*, Proposal Preparation Kit, February 1994.

10. See, for example, Brian Wright, "The Economics of Invention Incentives," *American Economic Review*, vol. 73 (September 1983), pp. 691–707; and Erich Kaufer, *The Economics of the Patent System* (Chur, Switzerland: Harwood Academic Publishers, 1988).

11. See Edwin Mansfield, "Patents and Invention: An Empirical Study," *Management Science*, vol. 32 (February 1986), pp. 173–81; Richard Levin and others, "Appropriating the Returns from Industrial Research and Development," *Brookings Papers on Economic Activity: Microeconomics, 1987*, pp. 783–820; and Wesley J. Cohen and others, "Appropriability Conditions and Why Firms Patent and Why They Do Not in the American Manufacturing Sector," working paper (Pittsburgh: Carnegie-Mellon University, June 1997).

12. But see F. M. Scherer, "The Innovation Lottery," New York University Law School conference paper, June 1998, which argues that individual inventors may be risk-averse but skewness-loving.

13. Edwin Mansfield and others, *The Production and Application of New Industrial Technology* (Norton, 1977), pp. 22–32.

14. Henry J. Grabowski and John Vernon, "A New Look at the Returns and Risks to Pharmaceutical R&D," *Management Science*, vol. 36 (July 1990), pp. 804–21.

15. Further analysis showed that the sample patents were quite different from run-of-the-mill invention patents. The average high-value sample patent was cited 15.7 times in subsequently issued U.S. patents, compared with only 6.02 times for all U.S. patents issued during the same period. Dietmar Harhoff and others, "Citation Frequency and the Value of Patented Inventions," *Review of Economics and Statistics* (forthcoming, 1999).

16. This is an important point, which is addressed in the discussion of high-technology venture capital funds.

17. Samuel Hollander, *The Sources of Increased Efficiency* (MIT Press, 1965), pp. 195–205.

18. See also Richard R. Nelson, "Uncertainty, Learning, and the Economics of Parallel Research and Development Projects," *Review of Economics and Statistics*, vol. 43 (November 1961), pp. 351–68.

19. Joseph A. DiMasi and others, "Research and Development Costs for New Drugs by Therapeutic Category," *PharmacoEconomics*, vol. 7 (February 1995), pp. 152–69. See also Joseph A. DiMasi, Henry Grabowski, and John Vernon, "R&D Costs, Innovative Output and Firm Size in the Pharmaceutical Industry," *International Journal of the Economics of Business*, vol. 2 (July 1995), pp. 201–19 (showing that larger firms tend to spend more per drug investigated, but less per drug approved, than smaller companies).

20. The lowest out-of-pocket cost, $1.5 million, was for a topical steroid (an ointment applied to the skin). The highest cost (not explicitly reported) was for an anti-infective; it was so high that it was excluded from group averages.

21. See Charles P. Himmelfarb and Bruce C. Petersen, "R&D and Internal Finance: A Panel Study of Small Firms in High-Technology Industries," *Review of Economics and Statistics*, vol. 76 (February 1994), pp. 38–51. Pharmaceuticals, with unusually high R&D/sales ratios, may be an excep-

tion. See F. M. Scherer, *Industry Structure, Strategy, and Public Policy* (HarperCollins, 1996), p. 388.

22. Himmelfarb and Petersen, "R&D and Internal Finance."

23. Bank of England, *The Financing of Technology-Based Small Firms* (London: October 1996), p. 41.

24. Ibid., p. 16. See also Guy Saint-Pierre, "The Globalization of Knowledge: R&D in Canada," *Technology in Society*, vol. 19 (1997, nos. 3/4), p. 272: "The long, successful, and highly respected Canadian banking business has always been marked by caution, and has traditionally preferred to loan on solid assets instead of intellectual capital." But at p. 278 he notes that Canadian banks were "becoming more open to supplying risk capital."

25. Bank of England, *The Financing of Technology-Based Small Firms*, p. 54.

26. See Anna Lee Saxenian, *Regional Advantage: Culture and Competition in Silicon Valley and Route 128* (Harvard University Press, 1994); and "The Valley of Money's Delight," Special Survey, The *Economist*, March 29, 1997.

27. It is drawn from Heidi Willmann, "Innovation in the Venture Capital Industry: A Study of American Research and Development Corporation," term paper, John F. Kennedy School of Government, Harvard University, May 1991. The summary here is drawn from F. M. Scherer, "The Size Distribution of Profits from Innovation," *Annales d'Economie et de Statistique* (1998, no. 49/50), pp. 502–03.

28. A fuller analysis appears in Scherer, "The Size Distribution of Profits from Innovation."

29. See George W. Fenn, Nellie Liang, and Stephen Prowse, "The Economics of the Private Equity Market," Federal Reserve Board staff study (Washington: December 1995), pp. 57–63. See also Venture Economics Inc., *Venture Capital Performance* (Boston: 1988), p. 94.

30. U.S. National Science Board, *Science & Engineering Indicators: 1998*, pp. 6-31 and A-381.

31. See "The Venture Capital Adventure," *Institutional Investor*, September 1981, p. 105; and "The Pension Funds Stoop to Conquer," *Business Week*, March 15, 1982, p. 138.

32. U.S. National Science Board, *Science & Engineering Indicators: 1998*, p. A-382.

33. See, for example, "'Angels Willing to Help Start-up Firms," *Boston Globe*, May 12, 1996, which reports an estimate that "angels" in the United States invest in 30,000 to 40,000 startup companies a year—most, presumably, not in high-technology fields.

34. U.S. National Science Board, *Science & Engineering Indicators: 1998*, pp. 6-31 and A-384.

35. See "The Boom in IPOs," *Business Week*, December 18, 1995, pp. 64–72.

36. There is some ambiguity because twenty-one of the sample's early IPOs did not achieve even infrequently traded status, and it is unclear whether the initial investments would actually have been made. Excluding those twenty-one, the initial $110,000 portfolio investment would have accumulated to $521,326 at the end of 1995, compared with $501,908 for investment in the NASDAQ index.

37. See "Going for the Golden Egg," The *Economist*, September 28, 1996, pp. 89–90; "Europe's Great Experiment," The *Economist*, June 12, 1998, pp. 67–68; and "The Euro's Warm-Up Act: IPOs," *Business Week*, June 22, 1998, p. 60.

38. U.S. National Science Board, *Science & Engineering Indicators: 1998*, p. 6-33, citing data reported in the yearbook of the European Venture Capital Association.

39. U.S. National Science Board, *Science & Engineering Indicators: 1998*, p. 6-33.

40. Bank of England, *The Financing of Technology-Based Small Firms*, p. 53.

41. The allusion is to David Landes's magisterial history of the industrial revolution, *The Unbound Prometheus: Technology Change and Industrial Development in Western Europe from 1750 to the Present* (Cambridge, U.K.: Cambridge University Press, 1969).

42. On industrial research subsidization, see notes 4–9 above.

43. Chancellor of the Duchy of Lancaster, *Realising Our Potential*, p. 51.

44. "Agreement on Subsidies and Countervailing Measures," part iv, article 8, reproduced in *Uruguay Round Trade Agreements, Texts of Agreements, Implementing Bill, Statement of Administrative Action, and Required Supporting Statements*, Message from the President of the United States to the Congress (Washington: September 27, 1994), p. 1541.

45. On the complexities for multinational corporations, see James R. Hines, "On the Sensitivity of R&D to Delicate Tax Changes: The Case of Multinationals," in Alberto Giovannini and others, eds., *International Taxation* (University of Chicago Press, 1993).

46. In the United States, companies have the option of capitalizing their R&D outlays and depreciating them over five years. Because losses can be carried forward for fifteen years, the capitalization option is rarely advantageous.

47. See, for example, "Venture Capitalists Wary of Tax Plan," *New York Times,* January 9, 1985.

48. See James M. Poterba, "Venture Capital and Capital Gains Taxation," in Lawrence H. Summers, ed., *Tax Policy and the Economy,* vol. 3 (MIT Press, 1989), pp. 47–67.

49. In the 1986 Tax Reform Act, the United States eliminated accelerated-depreciation (and investment tax credit) tax provisions previously in force.

50. See F. M. Scherer, "Demand-Pull and Technological Innovation: Schmookler Revisited," *Journal of Industrial Economics,* vol. 30 (March 1982), pp. 225–38.

51. See Edwin Mansfield, *Studies of Tax Policy, Innovation, and Patents: A Final Report* (University of Pennsylvania: October 1985), part III; U.S. Congress, Office of Technology Assessment, *The Effectiveness of Research and Experimentation Tax Credits* (Washington: 1995), pp. 44–50; and OECD, *Fiscal Measures to Promote R&D and Innovation,* DSTI report brief OCDE/GD(96)165 (Paris: 1996).

52. Mansfield, *Studies of Tax Policy,* p. 9.

53. This account is based on Bronwyn H. Hall, "R&D Tax Policy during the Eighties: Success or Failure?" in James M. Poterba, ed., *Tax Policy and the Economy,* vol. 7 (MIT Press, 1993), pp. 1–35.

54. Hall, "R&D Tax Policy during the Eighties." See also U.S. National Science Board, *Science & Engineering Indicators: 1998,* pp. 4-47 and 4-48.

55. See Donald G. McFetridge, "Intellectual Property Rights and the Location of Innovative Activity: The Canadian Experience with Compulsory Licensing of Patented Pharmaceuticals," paper presented at a National Bureau of Economic Research symposium (July 1997).

56. See Sandy Weisburst and F. M. Scherer, "Economic Effects of Strengthening Pharmaceutical Patent Protection in Italy," *International Review of Industrial Property and Copyright Law* (1995, no. 6), pp. 1009–24; and Jean O. Lanjouw, "The Introduction of Pharmaceutical Product Patents in India," Yale University Economic Growth Center Discussion Paper 775 (August 1997).

Chapter 6

1. Total population is taken as the denominator, in contrast to the labor force count used in figure 6-2, because for largely agrarian and homework-

oriented less-developed nations, it is difficult to know who is and who is not in the formal labor force.

2. U.S. National Science Foundation, *Research and Development in Industry: 1994* (Washington, D.C.: 1997), pp. 30–31.

3. U. S. National Science Board, *Science & Engineering Indicators: 1996*, p. 2-5.

4. U.S. National Science Board, *Science & Engineering Indicators: 1998*, pp. A-38 ff.

5. The Vietnam war injected erratic transients into the data, at first as students chose science and engineering curricula in the hope of avoiding conscription and then as randomized conscription depleted the stock of students.

6. One consequence of declining computer science enrollments was a severe shortage of programmers in the mid-1990s. See "Software Jobs Go Begging, Threatening Technology Boom," *New York Times*, January 13, 1998, pp. 1, D6; "Students Dropping Out for High-Tech Jobs," *International Herald-Tribune*, June 26, 1998; and "Is There Really a Techie Shortage?" *Business Week*, June 19, 1998, p. 93.

7. Data permitting a consistent extension beyond 1990 were not available.

8. Centre for Educational Research, Organisation for Economic Co-Operation and Development, *Education at a Glance: OECD Indicators* (Paris: 1995), p. 220.

9. See Engin Holmstrom and others, "Is There a Brain Drain from Science and Engineering?" *Issues in Science and Technology*, vol. 13 (Spring 1997), pp. 86–87.

10. Excluded from the Japanese totals are *Ronbun* doctorates awarded to industry employees on submission of a thesis to the universities at which they received baccalaureate degrees. Adding *Ronbun* degrees would approximately double the count of Japanese doctorates for 1992.

11. U.S. National Science Foundation, *Data Brief*, "Graduate Enrollment Drops for the Second Year in a Row," February 19, 1997; and *Data Brief*, "Doctorate Awards Increase in S&E Overall, but Computer Science Declines for First Time," November 7, 1997.

12. "The Best B Schools," *Business Week*, October 24, 1994, p. 65.

13. In Japan the problem of inadequate university education is remedied when large high-technology companies provide intensive on-the-job training for the young engineers and scientists they hire.

14. U.S. Bureau of the Census, *Statistical Abstract of the United States: 1996*, p. 17.

15. See "Supply Exceeds Demand for Ph.D.s in Many Science Fields," *New York Times*, July 4, 1995; and "U.S. Firms Experiment with Part-Time Scientists," *International Herald-Tribune*, January 31, 1997, p. 1.

16. On the problem of lagged educational response, see David M. Blank and George J. Stigler, *The Demand and Supply of Scientific Personnel* (New York: National Bureau of Economic Research, 1957); and Kenneth J. Arrow and William Capron, "Dynamic Shortages and Price Rises: The Engineer-Scientist Case," *Quarterly Journal of Economics*, vol. 73 (May 1959), pp. 292–308.

17. The most able scientists may also display the greatest interfield mobility. It is remarkable what a crucial role physicists (such as Erwin Schrödinger, Max Delbrück, Francis Crick, Maurice Wilkins, and George Gamow) played in working out how the structure of DNA determines heredity. See Susan Aldridge, *The Thread of Life: The Story of Genes and Genetic Engineering* (Cambridge University Press, 1996), pp. 21–35.

18. U.S. National Science Board, *Science & Engineering Indicators: 1996*, p. 3-12.

19. See "Up from the Wreckage of Russian Science," *Business Week*, October 27, 1997, pp. 153–54; and "Science in Russia: The Diamonds in the Rubble," *The Economist*, November 8, 1997, pp. 25–28.

20. Data for Taiwan, which were excluded from the United Nations tabulation, were added from the *Taiwan Statistical Data Book: 1996*. The data for Singapore were too incomplete for inclusion in a detailed analysis.

21. There is a bias in the statistics, since the less developed nations tend to have lower life expectancies and higher birth rates, so they have relatively more citizens in the university studies age category than do industrialized nations. Thus, if anything, the intensity of S&E study is even lower for the low-income nations relative to eighteen- to twenty-four-year-old cohorts.

22. Where INTENSITY is the number of S&E students per 100,000 population and GNP/CAP is GNP per capita, the following least-squares regression is obtained:

$$INTENSITY = 275.43 + 0.031 \text{ GNP/CAP}; r^2 = 0.273; n = 64.$$
$$(4.82)$$

The number in subscripted parentheses is a *t*-ratio. The coefficient 0.031 means that on average an increase of \$1,000 in income per capita leads to thirty-one more S&E students per 100,000 population.

Where DEGREES is the number of students enrolled in S&E degree-granting programs, the comparable regression is:

$$\text{DEGREES} = 249.32 + 0.020 \text{ GNP/CAP}; r^2 = 0.175; n = 58.$$
$$(3.49)$$

23. For the United Kingdom, the relevant number is 389,000; for Canada, 314,700.

24. The United Kingdom ranked twenty-fifth among the sixty-five nations, with 673 S&E students per 100,000 population. Peru was ranked fifth with 1,215 students according to the UN tabulations, but a check against data published in the *Statistical Abstract of Latin America*, vol. 31, suggested a sizable reporting error. Peru is therefore excluded from the ranking.

25. See, for example, "U.S. Funds Research at Former Soviet Germ Warfare Labs," *Boston Globe*, August 10, 1997, p. A2.

26. See, for example, "Forget the Huddled Masses: Send Nerds," *Business Week*, July 21, 1997, pp. 110–12.

27. For an older analysis, see U.S. National Academy of Engineering, *Foreign and Foreign-Born Engineers in the United States: Infusing Talent, Raising Issues* (Washington: National Academy Press, 1988).

28. "Non–U.S. Citizens Are 40 Percent of S&E Doctorate Recipients from U.S. Universities in 1995," National Science Foundation, *Data Brief*, August 19, 1996.

29. U.S. National Science Board, *Science & Technology Indicators: 1996*, p. 3-22.

30. Ibid., appendix table 3-15.

31. U.S. National Science Board, *Science & Engineering Indicators: 1998*, p. 3-20.

32. U.S. National Science Board, *Science & Engineering Indicators: 1996*, appendix table 3-14.

33. Ibid., p. 3-22.

34. See Walter Adams, ed., *The Brain Drain* (Macmillan, 1968).

35. U.S. National Science Board, *Science & Engineering Indicators: 1996*, p. 3-13.

36. Ibid., p. 2-6 and appendix table 2-5.

37. In 1993, females outnumbered males as teachers in U.S. public elementary and secondary schools by a ratio of 2.69 to 1. *Statistical Abstract of the United States: 1996*, p. 166.

38. Richard J. Murnane and others, *Who Will Teach? Policies That Matter* (Harvard University Press, 1991), pp. 35–36.

39. U.S. National Science Board, *Science & Engineering Indicators: 1998*, p. 1-24.

40. See figure 5-5.

Chapter 7

1. I ignore at some risk a related question—whether we will continue to be successful in containing the use of the massively destructive weapons also made possible by scientific and technological advances.

2. Derek J. de Solla Price, *Little Science, Big Science* (Columbia University Press, 1963).

3. The data are from Angus Maddison, *Monitoring the World Economy: 1820–1992* (Paris: OECD, 1995), pp. 194–97.

4. De Solla Price, *Little Science, Big Science*, p. 19.

5. See Derek C. Bok, *The Cost of Talent: How Executives and Professionals Are Paid and How It Affects America* (New York: Free Press, 1993); and Robert H. Frank and Philip J. Cook, *The Winner-Take-All Society* (New York: Free Press, 1995).

6. See Charles T. Clotfelter and Philip Cook, *Selling Hope: Lotteries in America* (Harvard University Press, 1989).

7. See, for example, Illinois Institute of Technology Research Institute, *Technology in Retrospect and Critical Events in Science* (the "TRACES" study) (Washington: National Science Foundation, 1968); and Edwin Mansfield, "Academic Research Underlying Industrial Innovations: Sources, Characteristics, and Financing," *Review of Economics and Statistics*, vol. 77 (February 1995), pp. 55–65.

8. U.S. National Science Foundation, *National Patterns of R&D Resources: 1996* (Washington: NSF Report 96-333, 1996), pp. 83–84. Price level changes are corrected using the gross domestic product price deflator. No adequate price deflator series exists for R&D alone. It is likely that the GDP deflator underadjusts for inflation, and hence real growth rates are overstated.

9. See James H. Billington, "Russia, Between a Dream and a Nightmare," *New York Times*, June 17, 1998, p. A31; and *The Face of Russia* (forthcoming).

10. See "Silicon Valley, PRC," The *Economist*, June 27, 1998, p. 65. In 1999 plans were being laid for a high technology–oriented NASDAQ–like stock market in China.

11. For case studies, see F. M. Scherer, *International High-Technology Competition* (Harvard University Press, 1992), pp. 52–102.

12. On the debate over its proposed extension in the United States, see "Higher Quota Urged for Immigrant Technology Workers," *New York Times*, February 23, 1998; Thomas L. Friedman, "Help Wanted," *New York Times*, April 14, 1998, p. A25; "Technology Worker Visa at Center of Industry Debate," *New York Times*, April 20, 1998, p. D10; Howard Gleckman, "High-Tech Talent: Don't Bolt the Golden Door," *Business Week*, March 16, 1998, p. 30; and "Bill to Increase Work Visas for Foreigners Gets New Lease on Life," *New York Times*, October 14, 1998, p. C3.

13. In 1996, 61 percent of foreign citizens receiving science or engineering Ph.D. degrees in the United States completed their studies without accumulated debt; for U.S. citizens, the no-debt fraction was 39 percent. U.S. National Science Foundation, Division of Science Resources Studies, *Issue Brief*, July 8, 1998.

Index

British–North American Committee

AT THE British–North American Committee's first meeting in New York City in December 1969, the following statement of its aims was authorized:

> The British–North American Committee has been established to study and comment upon the developing relationships between Britain, the United States, and Canada. It seeks to promote clearer understanding of the economic opportunities and problems facing the three countries, to explore areas of cooperation and possible friction, and to discover constructive responses. It believes that sound relations between these three countries in the context of an increasingly interrelated world are essential to future prosperity and seeks to promote better understanding through the collection of facts and their widespread dissemination.

During the 30 years since then, the British–North American Committee (BNAC) has broadly maintained these aims and has sponsored a series of some 48 objective studies undertaken by qualified experts in the three countries and published with the committee's approval. On the basis of these factual studies and of discussions at its meetings, the committee has also issued policy statements signed by its members.

BNAC's membership includes business, banking, labor, academic, and professional leaders from the United Kingdom, the United States, and Canada. It is sponsored by three nonprofit research organizations—the British–North American Research Association in London, the Center for Strategic and International Studies in Washington, D.C., and the C. D. Howe Institute in Toronto.

BNAC is a unique organization in terms of both its broadly diversified membership and the blending of factual studies and policy conclusions on issues important to the three countries it represents. It meets twice a year, once in the United Kingdom and once in North America. Its work is jointly financed by funds contributed from private sources in the United Kingdom, the United States, and Canada. Offices on behalf of the committee are maintained at 1800 K Street, N.W., Washington, D.C. 20006, and at Grosvenor Gardens House, 35-37 Grosvenor Gardens, London SW1W 0BS. Dr. Robin Niblett in Washington is the North American director of the committee, and Philip Connelly in London is the British director.

Sponsoring Organizations

The **British–North American Research Association** was inaugurated in December 1969. Its primary purpose is to sponsor research on British–North American economic relations in association with the British–North American Committee. Publications of the British–North American Research Association as well as publications of the British–North American Committee are available from the association's office, Grosvenor Gardens House, 35-37 Grosvenor Gardens, London SW1W 0BS (tel. 0171-828-6644). The association is registered as a charity and is governed by a council under the chairmanship of Sir Michael Bett.

The **Center for Strategic and International Studies** (CSIS) is a public policy research institution dedicated to analysis and policy impact. CSIS is the only institution of its kind that maintains resident experts on all the world's major geographical regions. It also covers key functional areas, such as international finance, U.S. domestic and economic policy, and U.S. foreign policy and national security issues. For more than three decades, the strategic approach of CSIS has emphasized long-range, anticipatory, and integrated thinking on a wide range of policy issues. The center's staff of 90 research specialists, 80 support staff, and 70 interns is committed to generating strategic analysis, analyzing policy options, exploring contingencies, and making recommendations.

Founded in 1962, CSIS is a private, tax-exempt institution. Its research is nonpartisan and nonproprietary. Sam Nunn chairs its Board of Trustees, and Richard Fairbanks is its president. CSIS offices are located at 1800 K Street, N.W., Washington, D.C. 20006 (telephone: 202/887-0200).

The **C. D. Howe Institute** is an independent, nonpartisan, nonprofit research and education institution. It carries out and makes public independent analyses and critiques of economic policy issues and translates scholarly research into choices for action by governments and the private sector. The institute was established in 1973 by the merger of the C. D. Howe Memorial Foundation and the Private Planning Association of Canada (PPAC).

While its focus is national and international, the institute recognizes that Canada is composed of regions, each of which may have a particular perspective on policy issues and different concepts of what should be national priorities. Participation in the institute's activities is encouraged from business, organized labor, trade associations, and the professions. Through objective examinations of different points of view, the institute seeks to increase public understanding of policy issues and to contribute to the public decision-making process. Thomas E. Kierans is president and treasurer. The institute's offices are located at 125 Adelaide Street East, Toronto, Ontario M5C 1L7 (telephone: 416/865-1904).

BNAC members

Following are BNAC members representing three countries.

British Members

Scott Bell
Group Managing Director
Standard Life

Sir Michael Bett*
Chairman, Cellnet and
First Civil Service Commissioner
Office of the Civil Service
    Commission

Martin Broughton
Chairman
British American Tobacco

Roger Carr
Chairman
Thames Water

Sir Anthony Cleaver*
Chairman
AEA Technology

Bill Cockburn
Group Managing Director, BT UK
BT

Sir Frederick Crawford*
Chairman
Criminal Cases Review
    Commission

Sir John Daniel*
Vice Chancellor
The Open University

Baroness Dean of
    Thornton-le-Fylde
Chairman
The Housing Commission

Sir Richard Evans
Chairman
British Aerospace

Niall FitzGerald
Chairman
Unilever

Sir Patrick Gillam
Chairman
Standard Chartered Bank

Sir Angus Grossart
Vice Chairman
Royal Bank of Scotland

Dr. Nigel Horne
Chairman
Alcatel UK

Sir John Kingman
Vice Chancellor
University of Bristol

Sir Sydney Lipworth
Chairman
Zeneca Group

Roger Lyons
General Secretary
MSF

George Mallinckrodt, KBE
President
Schroders

Lord Marshall
Chairman
British Airways

Charles Miller Smith
Chairman
ICI

John Monks
General Secretary
Trades Union Congress

Mark Moody-Stuart
Chairman
Royal Dutch/Shell Group of
    Companies

Sir Geoffrey Mulcahy
Chief Executive
Kingfisher

Keith Orrell-Jones
Chairman
Smiths Industries

Larry G. Pillard
Chief Executive
Tate & Lyle

Sir Brian Pitman*
Chairman
Lloyds TSB Group

George Poste
Director and
Chief Science and Technology
    Officer
SmithKline Beecham

Sir Ian Prosser*
Chairman & Chief Executive
Bass

Sir Bob Reid*
Deputy Governor
Bank of Scotland

James H. Ross
Chairman
The Littlewoods Organization

Alan Rudge
Chairman
W. S. Atkins

Peter Salsbury
Chief Executive
Marks & Spencer

Mrs. Steve Shirley
Life President
F. I. Group

Lord Simpson
Chief Executive
General Electric Company

Jonathan Taylor*
Chairman
Ellis & Everard

Hon. Barbara S. Thomas
Executive Chairman
Whitworths Group

Ed Wallis
Chairman
PowerGen

Derek Wanless
Director, Group Chief Executive
Natwest Group

Simon Webley
Research Director
Institute of Business Ethics

Viscount Weir*
Chairman
BICC

Humphrey Wood
Chairman
Vitec Group

U.S. Members

George F. Becker
International President
United Steelworkers of America

Geoffrey Bell
President
Geoffrey Bell & Company

John C. Blythe
President
Foster Wheeler International
    Corporation

William Bradford*
Chairman and CEO
Dresser Industries

Richard C. Breeden
President and CEO
Richard C. Breeden & Company

Richard M. Clarke
Chairman
Yankelovich Partners

James D. Ericson
President and CEO
Northwestern Mutual Life
    Insurance Company

Edward A. Feigenbaum
Professor of Computer Science
Stanford University

William B. Finneran
Chairman
Edison Control Corporation

Peter M. Gottsegen
Partner
CAI Advisors & Company

Alan R. Griffith
Vice Chairman
Bank of New York

Hon. Alexander M. Haig Jr.
Chairman and President
Worldwide Associates

Hon. John G. Heimann*
Chairman, Financial Stability
  Institute
Bank for International Settlements

James M. Hoak
Chairman
Hoak Capital Corporation

Donald P. Jacobs
Dean
J. L. Kellogg Graduate School of
  Management
Northwestern University

John T. Joyce
President
International Union of Bricklayers
  and Allied Craftworkers

David Levy, M.D.
Chairman and CEO
Franklin Health

Edward E. Madden
Vice Chairman
Fidelity Investments

J. R. Massey
President
Mobil Europe and Africa

David E. McKinney
Executive Secretary
Thomas J. Watson Foundation

Richard L. Measelle*
Managing Partner
Arthur Andersen

James E. Perrella*
Chairman, President, and CEO
Ingersoll-Rand Company

Lee R. Raymond
Chairman and CEO
Exxon Corporation

David A. Reed
Senior Vice Chair—Global/Major
  Accounts, Industries, Sales, and
  Marketing
Ernst & Young

Robert D. Rogers*
President and CEO
Texas Industries

Patrick G. Ryan
President and CEO
Aon Corporation

Frederic V. Salerno
Senior Executive Vice President,
  and CFO
Bell Atlantic Corporation

Thomas R. Saylor*
Chairman and CEO
Cambridge Biomedica

F. M. Scherer
Aetna Professor of Public Policy
and Management
John F. Kennedy School of
Government
Harvard University

James Schiro
CEO
PriceWaterhouseCoopers

Hon. James R. Schlesinger
Senior Advisor
Lehman Brothers

Brian E. Stern*
President, Xerox New Enterprises
Xerox Corporation

Jan H. Suwinski
Professor of Business Operations
Samuel Curtis Johnson Graduate
School of Management
Cornell University

Frederick B. Whittemore
Advisory Director
Morgan Stanley & Company

Kern Wildenthal, M.D.
President
University of Texas

Margaret S. Wilson
Chairman and CEO
Scarbroughs

Canadian Members

A. Charles Baillie
Chairman and CEO
Toronto Dominion Bank

John E. Cleghorn
Chairman and CEO
Royal Bank of Canada

Marshall A. Cohen*
Counsel
Cassels, Brock & Blackwell

Léon Courville
President, Personal and
Commercial Banking, and COO
National Bank of Canada

Jean Claude Delorme
Corporate Director and Consultant

Hon. J. Trevor Eyton*
Senior Chairman
EdperBrascan Corporation

Maureen Farrow
Economap Inc.

Peter C. Godsoe*
Chairman and CEO
Bank of Nova Scotia

Kerry L. Hawkins
President
Cargill Limited

G. R. Heffernan
President
G. R. Heffernan and Associates

Thomas E. Kierans
President and CEO
C. D. Howe Institute

Michael M. Koerner
President
Canada Overseas Investments

Jacques Lamarre
President and CEO
SNC LAVALIN Group

Claude Lamoureux
President and CEO
Ontario Teachers' Pension Plan
    Board

Pierre Lortie
President and COO
Bombardier International

William A. Macdonald
President
W. A. Macdonald Associates

Hon. Roy MacLaren, P.C.
High Commissioner
Canadian High Commission

John D. McNeil*
Chairman
Sun Life Assurance Company of
    Canada

Ronald Osborne
President and CEO
Ontario Power Generation

Robert B. Peterson
Chairman, President and CEO
Imperial Oil Limited

Alfred Powis
Director
Noranda Inc.

J. Robert S. Prichard
President
University of Toronto

Sheldon M. Rankin
President and CEO
J & H Marsh & McLennan Limited

Arthur R. A. Scace*
Chairman
McCarthy Tétrault, Barristers &
    Solicitors

Dr. Michael D. Sopko
Chairman and CEO
Inco Limited

James Stanford
President and CEO
PetroCanada

Thomas H. B. Symons*
Founding President and Vanier
    Professor Emeritus at Peter
    Robinson College
Trent University

William I. M. Turner Jr.*
Chairman and CEO
William and Nancy Turner
    Foundation

Colin D. Watson
President and CEO
Spar Aerospace Limited

*Executive Committee Member